SCIENCE AND ITS CRITICS

JOHN PASSMORE

Science and technology have been thought of as either salvation or scourge. Bacon and Descartes believed that scientists discover the causes and secret motions of things, while closer to our own time, Friedrich Nietzsche felt that science belittles man. Modern critics of science claim that science explains a mechanism, not a purpose—that science asks "how," not "why." Whether this fear that science might depersonalize human beings has any validity is one of the leading philosophical issues of our time.

Passmore establishes science as a legitimate intellectual pursuit as he answers science's critics, carefully presenting the claims for the understanding and knowledge derived from free inquiry. Science, he argues, can demolish myths about the degree of autonomy human beings possess without weakening the everyday concepts of freedom, respect, and dignity. He recognizes that science can be dangerously optimistic in its unilinear solutions to social problems, and it can be structured so as to emphasize analytic procedures even

in cases wh
accumulati
urges that
in favor of
natural history, for it is through these abstract and analytic investigations that we have learned to understand and control the world around us.

Science, like other pursuits, has its toilers as well as its "aristo-scientists"—the science elite. The threat from aristoscience, Passmore recognizes, comes from the desire to confront a problem purely in its own terms. Thus the aristoscientist studies cancer in cells, ignoring the environmental and public health aspects of the disease. Yet it is the province of social planners, or the government, to make the connection between the individual findings and the social effects of investigations.

Despite the strictures levelled against science, Passmore responds that "it does not destroy uniqueness; it is not hostile to the imagination; it does not falsify by being abstract; it is as objective as the human condition permits. . . ."

This is an eloquently argued case for a proper discernment of the place of science in modern life in which Passmore takes on Marxist critics, Theodore Roszak, Jacques Barzun, Barry Commoner, and others.

John Passmore is in the Department of Philosophy at Australia National University, Canberra. He has lectured at many universities in Great Britain, Germany, Japan, and the United States.

Science
and Its
Critics

Science and Its Critics

John Passmore

Mason Welch Gross Lectureship Series

RUTGERS UNIVERSITY PRESS
New Brunswick, New Jersey

Library of Congress Cataloging in Publication Data

Passmore, John Arthur.
Science and its critics.

(Mason Welch Gross lectureship series)
1. Science—Philosophy. I. Title. II. Series.
Q175.P3424 501 77-12049
ISBN 0-8135-0852-5

Manufactured in the United States of America

Contents

Acknowledgments

This book is associated in my mind with an exce
number of pleasant memories. It grew out of th
sohn Lecture delivered at Monash University ur
Revolt against Science" and in the presence of
published in *Search* 3(1972): 415–22; paragrar
porated in the present text with the permis
and New Zealand Association for the A
From that point on it gradually expande
first lecture was delivered at the Universit
Memorial Lecture and at the University of He
as a whole was first tried out in the Programme on
nology and Society at Cornell University, under
table, but by no means uncritical, eye of Professor Max Blac
then, in summary, with a group of students and staff on a lawn on
tropic night in Honolulu. Revised, it was delivered in the John Curtin Medical School of the Australian National University under the auspices of its Committee on Science and Technology Policy Research. And then, finally, as the Mason Gross Lectures at Rutgers University. In each case I learned much from the discussion which followed the lecture and greatly enjoyed the hospitality which reached its peak, from an already extraordinarily high level, at Rutgers University at the hands of Dr. Gross himself and Professor Richard Schlatter.

It will be apparent from their history that the lectures were not at all designed for audiences consisting wholly of professional philosophers. I am very conscious of the fact that I have sometimes skirted around, or skated over, deep philosophical problems. But I hope, all the same, that I have written a philosophical book, even if one which is accessible to a wider audience than the closer technical studies which naturally provide philosophy with its center. I should add that the printed version of the lectures differs considerably from what was anywhere delivered.

My thanks are due to Miss Isabel Sheaffe for her patience with the multiple retypings which the composition of this book has called for; to Mr. David Dumaresq for assistance in coping with the wide-ranging literature and with the checking of detail; to my wife for a great deal of careful reading and criticizing of the text at various stages in its composition. And once more, to my old teacher, John Anderson, who would have disagreed with much which I have here written but whose influence, nevertheless, is everywhere present.

John Passmore

The scope of the practical control of nature newly put into our hand by scientific ways of thinking vastly exceeds the scope of the old control grounded in common-sense. Its rate of increase accelerates so that no one can trace the limit; one may even fear that the *being* of man may be crushed by his own powers. . . . He may drown in his wealth, like a child in his bath tub, who has turned on the water and cannot turn it off.

William James

Chapter 1

Antiscience and Scientific Explanations

On the face of it, science has been one of the more successful forms of human enterprise. Consider the ambitions of science as Bacon and Descartes proclaimed them at the beginning of the modern scientific era. The scientists of Bacon's *New Atlantis* set out to acquire "knowledge of causes, and secret motions of things; and the enlarging of the bounds of human empire, to the effecting of all things possible." Descartes was no less hopeful. By conjoining the artisan's skills with the philosopher's intellect science would eventually generate, so Descartes anticipated, an "infinity of arts and crafts, enabling us to enjoy without any trouble the fruits of the earth and the good things that are found there."[1]

Surely, one might well feel inclined to assume, these ambitions have come to fruition. An ordinary household refrigerator is sufficient evidence of that fact. Making use of the physicist's theory of heat, his understanding of the "causes and secret motions of things," engineers and designers have produced a machine, a method of storage, which enables us, quite literally, to "enjoy the fruits of the earth and of the good things that are found there." And this with, anyhow, a diminishing minimum of trouble. If there is still much to be learned, much to be done, the path, on this view, lies clear before us—more science, more technology so that we shall have *total* understanding of everything that exists, *total* control

1. Francis Bacon, *New Atlantis* in *Works*, ed. J. Spedding, R. L. Ellis, and D. D. Heath, 14 vols. (London: Longman, 1857–74), 3:156; Rene Descartes, *Discourse on Method,* book 6.

over our environment, making of everything in it an instrument for the fulfillment of human desires.

From the earliest days, however, such vast ambitions, theoretical and practical, have aroused a measure of unease. Antiscience is almost as old as science. We can recognize it, in relation to a not yet technology-oriented Greek science, in the dialogues of Plato and the comedies of Aristophanes; in relation to modern science, we encounter it, well before our own time, in such philosophers as Berkeley, such satirists as Swift, such poets as Blake, such polymaths as Goethe. And this is only to mention great figures, to say nothing of a motley array of popular preachers and science-horror novelists. When Mary Shelley wrote *Frankenstein*, an appallingly bad novel which has nevertheless seized the Western imagination, science-based technology was scarcely under way.

As the names we cited will suggest, antiscience has taken a wide variety of forms. At one extreme, as in William Blake, it is directed against science as a whole—"I come," he once pronounced, "to cast off Bacon, Locke and Newton." Not only is antiscience directed against science, indeed, but against the abstracting, analytic intellect, in the supposed interests of the intuitive imagination. At the opposite extreme, then preferring to describe itself as "antiscientism" rather than as "antiscience," its objection, so its proponents tell us, is not to science as such but only to any attempt to extend the reign of science beyond its proper, very limited, sphere. Kierkegaard was happy to let science deal with plants and animals and stars. "But to handle the spirit of man in such a fashion," he wrote, "is blasphemy."[2]

Science, so Kierkegaard is suggesting, should stop short at the human being; the human spirit lies beyond its scope. Metaphysicians have sometimes drawn the boundary at a rather different point. Science, they have argued, is quite unobjectionable provided that it does not pretend to touch upon really fundamental matters, provided that it confines itself to the ordinary practical affairs of life. Unlike Kierkegaard, such metaphysicians—F. H. Bradley is a case in point—will permit science to investigate everything, even human beings, but only with this proviso: that, renouncing the overweening ambitions of a Bacon, it makes no pretence to discover ultimate causes.

2. Søren Kierkegaard, *Journals*, ed. and trans. Alexander Dru (London: Oxford University Press, 1938), p. 182.

Many great scientists have shared this limited degree of scepticism; it would be absurd to classify them as antiscientists. But let us borrow a phrase from the Marxists: such scientists were "objectively," although not at all "subjectively," antiscientists. In intention—"subjectively"—they sought no more than to confine science within limited bounds; their effect—"objectively"—has been to lend encouragement to the more virulent, deeply engaged, antiscientists. That is why we shall have to pay so much attention to their arguments.

To this situation, there are many parallels. Convinced Christians have formulated powerful criticisms of Christian theology and the Christian church, criticisms which have been taken over by anti-Christians; philosophers have criticised their metaphysically ambitious fellow philosophers in terms which have lent support to antiphilosophers. By the very nature of the case, those who denounce theology, science, or philosophy as a whole and those who renounce, only, its loftier ambitions will share many arguments in common.

With this preliminary explanation, let us begin to look more closely at the Bacon-Cartesian claim that the scientist discovers the "causes and secret motions of things." This has been questioned, by scientists and antiscientists alike, on a wide variety of grounds. (The views I shall be considering are by no means so popular with philosophers of science as they once were. But on the continent of Europe they still have strong support. And in Anglo-American philosophy of science they were predominant for more than half a century, declining only in the 1950s. One need not be surprised that antiscientists like Roszak still presume them to be orthodox.)

Science, their argument runs, can at most describe what goes with what, *how* things happen as distinct from *why* they happen, and therefore knows nothing of "causes and secret motions." This, one might at first think, is a wholly implausible line of criticism. For when we want to know why things happen as they do—why, to take elementary examples, the tides rise and fall or why some men have heart attacks and others do not—is it not precisely to science that we turn? Sometimes, of course, science fails us. Sometimes it has to confess to ignorance, sometimes it can tell us that such-and-such happens—that, let us say, aspirin relieves pain—but not *why* it does so. But to describe such cases as "scientific failures" is implicitly to recognize that science *often does* explain why. It is as compared with science's successes that we count these as failures. And who else, if not the scientist, has any chance of converting such failures into successes?

Scientific answers to our "why" questions, the reply would however come, only tell us "how"—how, let us say, the movements of the moon are related to the rise and fall of the tides—in a manner which is ideally representable as a set of equations. This does not explain *why* the moon exerts a gravitational pull. And in so far as science does attempt to answer such a question, any answer it has to offer would still take the form of a further "how." Abstract and highly generalized as they no doubt sometimes are, scientific explanations are still, on this view, brute facts. And to refer to a brute fact never, so it is said, really tells us why. This line of attack is by no means confined to metaphysicians and theologians. "These days everybody tells us *how*," complains a character in a recent Japanese film, "but nobody tells us why." And in a science fiction novel, the situation is summed up thus: "We have transferred our faith to science—the explanation for everything which explains nothing."

What can such critics have in mind? What, if not what science does by creating and testing theories, would count as "telling us why?" If scientific explanations were *really* satisfactory, if they *really* told us why, then—so it has often been replied—first, they would refer us to reasons so fundamental that we should at once see why everything must be as it is; secondly, these fundamental reasons would take the form of purposes; thirdly, they would explain everything, leaving no room whatsoever for chance or coincidence. All that science does, in contrast, is to tell us that one thing happens as a result of something else happening; it draws attention to a mechanism, not a purpose; it sets aside much that happens as, by the nature of the case, not admitting of explanation. So it leaves us, if we are in search of genuine explanations, as dissatisfied as ever.

Most typically, this objection comes from metaphysicians. A modern defender of Hegel, J. N. Findlay, tells us that facts need to be grounded, as science does not ground them, in something that is "both *self*-explanatory and *all*-explanatory." Hegel's appeal, he goes on to add, is to those who find "something logically absurd, imperatively requiring supplementation, in any mere fact, or collocation of mere facts, or even mere collocation of necessities."[3] And who are prepared, we might add, to dismiss Newton's experiments on colors, in Hegel's manner, as "complicated, bad, pettily done, mean and dirty," because they do not point towards such ultimate explanations.[4] But it is not only metaphysical absolutists who try to persuade us that science fails to explain. David Hume, certainly no absolutist, wrote thus of Newton: "While Newton seemed to draw off the veil from some of the mysteries of nature, he showed at the same time the imperfections of the mechanical philosophy; and thereby restored her ultimate secrets to *that obscurity, in which they ever did and ever will remain*."[5] The difference, of course, is that the metaphysical absolutist believes that he, although not the scientist, can discover the true causes and secret motions of things; the sceptical Hume denies that *anyone* can do so. Nature's "ultimate secrets," so he is suggesting, lie beyond human ken. And Hume found disciples, surprisingly enough, among scientists.

What could have induced scientists, of all people, to adhere to a philosophy which, like Hume's, is avowedly sceptical? Their motives were various. The most devoted disciples of Hume were the nineteenth-century agnostics; one of them, Thomas Huxley, wrote a book on him. What they disliked above all else was metaphysics, especially metaphysics masquerading as a "philosophy of nature," a competitor to science. In Hume's manner, the agnostics did not deny that there were "secret motions"; what they did deny was that anyone, scientist or metaphysician, could possibly uncover them. They were prepared to surrender the claim that science could do so if the same arguments disqualified their rivals—the metaphysicians, the theologians. Since scientists were obviously better than their critics at "describing how," that left them in command of the field.

Huxley's agnosticism is sufficiently familiar. But there were distinct traces of agnosticism, too, in that distinguished French scientist and philosopher of science, Henri Poincaré, and in a form

3. J. N. Findlay, foreword to G. W. F. Hegel, *Philosophy of Nature*, trans. A. V. Miller (Oxford: Clarendon Press, 1970), pp. ix, x.

4. Hegel, *Philosophy of Nature*, p. 211.

5. David Hume, *History of England*, 8 vols. (Oxford: Talboys, 1826), 8:296. The italics are mine.

which was to be very influential. All the scientist can hope to do, Poincaré argues, is to discover relationships. When the scientist asserts that such relationships hold between, let us say, electrons, "these are merely names of the images we have substituted for the real objects which Nature will hide for ever from our eyes."[6] If Poincaré is right, then the most that scientists can discover is that certain types of relationship hold. As it was later to be put: science is about form, not about content. What these relationships hold between and, hence, what are the secret motions in nature, is something it lies quite beyond science's competence to discover.

Agnosticism, however, is a distinctly odd doctrine, with its central thesis that we can somehow be completely confident that *there are* secret motions and causes without ever being in a position to know *what* they are. Some scientists, positivists like W. K. Clifford, drew attention to this fact; it is absurd, they argued, to complain that scientists must remain ignorant of "real causes" when we have no reason whatsoever for believing that any such "real causes" exist at all. How can we possibly know that nature is hiding something from us which we shall *never* discover?

The best solution, such positivists concluded, is to drop "why" questions altogether, as not only pointless but even meaningless. It is not that we cannot answer them, as the result of some human limitation. Rather, and in this quite unlike "What precisely does happen?" the question "Why does that happen?" does not raise a genuine issue. Its syntax deceives us; it is not asking a question.

Agnostics and positivists agreed, however, on the crucial point. It is not the task of science, or of any other discipline, to look for causes and secret motions. The most the scientist can do is to summarize in a convenient form the experiences he has had in his laboratory, if in a form which would at the same time—although the positivists found it hard to explain how—offer guidance about the future.

It would no doubt be absurd simply to attribute such a view to antiscience. Indeed, it has often been condemned as science in its most arrogant mood—"scientism"—trying to persuade us that the kind of description at which the scientist is adept represents all that

6. Henri Poincaré, *Science and Hypothesis* (1905; reprint ed., New York: Dover, 1952), p. 161.

human beings can hope to achieve in the way of understanding. But it did have the effect of converting science into a rather dull pursuit of useful summaries. Mathematical physics, so Poincaré tells us, draws up a catalog of the facts discovered by experiment.[7] And useful though catalogs are, exciting is scarcely the word we should naturally apply to them. Not surprisingly, so restricted a view of science left many scientists restless, inclined to believe that beyond science there must lie some other route to knowledge, not merely descriptive, but explanatory.

Pierre Duhem, distinguished at once as a physicist, a philosopher, and a historian of science, fully accepted the positivist account of science, although he presented it in a far more sophisticated and mathematically conscious form. But the limitations of science, so he argued against the positivist and agnostic, are not the limitations of human thinking; we can find *real* explanations outside science, in metaphysics, in religion. When—a convinced Roman Catholic—he was accused of importing theology into science, of presenting "the physics of a believer," he rightly replied that his account of scientific knowledge could be, and was, accepted by the positivist or the agnostic nonbeliever as much as by a believer like himself, that he had by no means invented the view that "a physical theory is not an explanation."[8] And Duhem's is the form in which scientific positivism has been picked up by such professional antiscientists as Theodore Roszak and used as a stick to beat science with. "Without a philosophical foundation in metaphysics," he takes Duhem to be arguing, "science could never be anything more than a kind of elite technology."[9] It could not, that is, offer us an understanding of the world as distinct from the power to control it. Just because they lend themselves to such an interpretation, positivism and agnosticism are "objectively" antiscience, however unwittingly.

But are they right? What Duhem and the positivists argued is no doubt true, up to a point, of the sort of theoretical physics they had in mind. The elementary propositions of Galilean mechanics which relate the velocity of a falling object to the distance it has

7. Ibid., p. 144.

8. Pierre Duhem, *The Aim and Structure of Physical Theory*, trans. P. P. Wiener (Princeton: Princeton University Press, 1954), p. 19.

9. Theodore Roszak, ed., *Sources* (New York: Harper and Row, 1972), p. 79.

fallen, its initial velocity and acceleration due to gravity, do not by themselves explain anything; neither why the body falls nor why it stops falling when it does. They take the form of equations; they do not contain that distinction between ground and consequent which is essential to explanations. The fact remains, however, that even such formulae are *intimately related to* explanation, in two distinct ways. They rule out a very plausible-looking explanation of why a stone falls faster from a cliff than does a feather, namely that the stone and the feather differ in mass. For mass plays no part in the equations. We do not have to take it into account in our calculations. And they can be *used to explain* why, to take the simplest possible example, an acorn dropping from a tree hits us with an exceptional force. It must either have had a high initial velocity—as a result, perhaps, of someone having thrown it down—or else it must have dropped from the very top of the tree. And not only the force of falling acorns but all kinds of complex movements, some of them distinctly esoteric, can be thus explained with the help of elementary mechanics.

Something similar is true of the equations of modern physics. Merely to assert that mass and energy are mathematically related does not of itself explain anything. But it can be used to explain a great deal. So although one can see why the positivists insisted that the propositions of mathematical physics do not explain, it does not follow that *science* does not explain—unless, as some philosophers of science appeared to do, we identify science with the set of equations belonging to mathematical physics. Furthermore, science explains by discovering "causes and secret motions": so—to take examples from outside physics—bacteria, viruses, osmosis, and genetic mutations are "causes and secret motions" which biologists have discovered and now use to explain the workings of our bodies in sickness and in health. Once they are discovered, of course, they are no longer "secret": but they *were* "secret," and they had to be discovered. And they are not fictions, not "images"; to learn about them is to learn something about the content of the world. The same is true of electrons.

No doubt, in explaining *why*, the scientists are still telling us *how* something happens. But that is only to say that a "why" question can be answered by a "how" statement, why John Smith's heart fails by referring to *how* he eats and *how* what he eats affects his arteries. In other words, the supposed antithesis between explaining *why* and telling *how* has no real force. Or, more accurately, its force relates only to a specific context. We want to know not *how* John Smith died, this we already know, but *why* he died. But this death is explained as a consequence of *how* he lived.[10]

As for the absolutist metaphysicians, their supposedly self-explanatory and all-explanatory absolutes conspicuously fail to explain. To explain, in the sense that is now relevant, is to make it plain why things happen in one way rather than another; why, let us say, planets move in elliptical rather than circular courses. But the so-called self-explanatory metaphysical absolutes simply cannot do this; it is characteristic of such absolutes that they are compatible with things happening in any way whatsoever. What professes to explain everything can in fact explain nothing, as Hume long ago pointed out. If this were not so, then the absolutist "explanations," like any other explanations, could be overthrown by new scientific discoveries, contradicting the doctrines that were allegedly deducible from them. And this possibility the absolutist will not allow. (Compare Wittgenstein's observation: "*How* things are in the world is a matter of complete indifference for what is higher.")[11]

But even if we insist—against positivists, agnostics and metaphysicians alike—that science, in a wide range of cases, can and does explain why things happen in one way rather than another, we are still left with the objection that science calls upon us to accept brute facts, that it cannot explain "why there is anything at all." This is the ground on which science, at least in its twentieth-century form, often finds itself dismissed, as by Heidegger, as a "technical practical business of gaining and transmitting information," incapable of awakening, and in fact emasculating, the spirit of genuine inquiry.[12] It is the spirit, too, in which Findlay suggests that the mind cannot rest "in any mere fact, or collocation of mere facts, or even mere collocation of necessities."

10. For a further discussion of explanation see John Passmore, "Explanation in Everyday Life, Science and History," *History and Theory* 2:105–23; this article is reprinted in *Studies in the Philosophy of History*, ed. George H. Nadel (New York: Harper and Row, 1965).

11. Ludwig Wittgenstein, *Tractatus Logico-Philosophicus*, trans. D. F. Pears and B. F. McGuiness (London: Routledge and Kegan Paul, 1961), 6.432.

12. See for example Martin Heidegger, *Introduction to Metaphysics*, trans. R. Manheim (New Haven: Yale University Press, 1959), p. 49.

The reply to such criticism, as I have already hinted, is that the mind has no option but so to rest. However far we go in our investigations we shall always come to such a fact or such a collocation. Any answer to "Why is there anything at all?" will have to be of the form "Because there is so-and-so," an answer which still permits us to raise the question why there is *it*. Of course, metaphysicians have tried to avoid this regress by setting up the concept of a self-explanatory being, whose nature is such that the question "Why does it exist?" has no meaning. It cannot but exist. But, in my judgment, they have failed to show—I obviously cannot pause to argue the point in detail—either that there could be any such being or that, if there were, it could in any way explain the existence of anything else.

So even if we reject Clifford's view that every question of the form "Why does such and such happen?" fails to raise a genuine issue, we might still wish to maintain that "Why does *anything* happen?" is in precisely this position. But we should be foolish to deny that not only Heidegger but very many people have found the question, "Why is there anything at all?" a deeply disturbing one and have rejected science as "superficial" precisely because it does not attempt to answer it. To set the question aside as not raising a genuine problem involves a degree of intellectual asceticism, a kind of Stoicism, in the popular sense of that word. Such an asceticism, such a Stoicism, is quite extensively called upon, so we shall find, if we restrict ourselves to science-type explanations. And a refusal to accept it lies at the heart of some of the deepest, most far-reaching, forms of antiscience.

When it is said, however, that science fails to tell us "why," what the critic often has in mind is not that the explanations which science offers fail to get us beyond the level of brute facts to the level of ultimate metaphysical explanation, but rather that they fail to tell us what we really want to know—the purpose which lies behind what happens.

That is the ground on which Plato represents Socrates as objecting to the Greek science of his day. Socrates, Plato tells us, went to Anaxagoras in search of wisdom; Anaxagoras, so Socrates hoped, would first of all tell him whether the earth was round or flat and would then go on to explain why this *must* be so, meaning by "must" why it *is best that it should be so*. Instead, Anaxagoras talked only about physical causes. And Socrates goes on to draw an analogy with his own situation, awaiting execution. It is as if, he says, someone were to explain why he had not run away from his prosecutors by talking about his muscles and his bones while never mentioning, what is the crucial point, that he *thought it best* to stay and face the charges brought against him.

The purposive, teleological, mode of explanation dominated the mediaeval world; the history of science has been the history of a long fight against it. But the demand for such explanations still persists. Consider Ursula Brangwen in D. H. Lawrence's *The Rainbow*, meditating in her laboratory: "But the purpose, what was the purpose? Electricity had no soul, light and heat had no soul. Was she herself an impersonal force, or conjunction of forces, like one of those? . . . For what purpose were the incalculable physical and chemical activities nodalised in this shadowy, moving speck under her microscope? What was the will that nodalised them and created the one thing she saw? What was its intention?" Notice the movement of Ursula Brangwen's thought: "Electricity had no soul . . . Was she herself an impersonal force, like one of those?" And then the leap to: "What was the will that nodalised these physical and chemical activities into a living organism?" Undoubtedly, human beings would feel in certain respects more "at home" in a universe in which they could sensibly ask in respect to everything that happens: "What did that happen for?" or in respect to anything which exists: "What is it for?"—where "What is it for?" is equivalent to "What *human* interest does it serve?" The biologist Marston Bates still finds ground for complaining that when people come upon an unfamiliar species, they immediately ask "What good is it?" and mean "In what respects is its existence helpful to human beings?"[13] They assume, that is, that whatever is must be for human use. But at least species *can be* humanly useful. In contrast, such questions as "Why is it better for the planets to move in ellipses than in circles?" we should now set aside as absurd, as asking a question which, in relation to planets, has no application.

Our feeling of "homelessness" in a world which, unlike human artifacts, is not made for us, gives its impetus to a great deal of

13. Marston Bates, *The Forest and the Sea* (London: Museum Press, 1961), p. 12.

antiscientific metaphysics, with its object to make of the universe something in which men and women will feel "at home," a phrase Hegel explicitly uses, as distinct from our merely being "housed" in it. Philosophy has often tried to satisfy that desire by suggesting that natural processes, appearances to the contrary notwithstanding, express an intention, exhibit a purpose. "Philosophy is really homesickness," wrote Novalis, "it is the urge to be at home everywhere."[14] We need feel no surprise that science, in opposing that attitude to the world, should still meet with resistance. And this is especially so because behind the resistance to science's claim to explain there lies a deeper fear, the fear that science might eventually demonstrate that the language of intentions, of purposes, is as inappropriate as a description of human behavior as it is of inanimate behavior: that science might end up by deanthropomorphizing human beings, paradoxical as this might sound, by making it foolish to ask even in relation to human behavior "What did he do that for?" as distinct from "What, in him, gave rise to that action?"

"Has there not been since the time of Copernicus," Nietzsche asks, "an unbroken progress in the self-belittling of man? . . . Alas, his belief in his dignity, his uniqueness, his irreplaceableness in the scheme of existence, is gone—he has become animal, literal, unqualified, and unmitigated animal, he who in his earlier belief was almost God. . . . *All* science . . . nowadays sets out to talk man out of his present opinion of himself, as though that opinion had been nothing but a bizarre piece of conceit."[15] In B. F. Skinner's *Beyond Freedom and Dignity* Nietzsche would find the fullest realization of what so disquieted him. "The traditional concept of man," Skinner writes, "is flattering; it confers reinforcing privileges."[16] And it is precisely that traditional flattering, privileged, position which Skinner is trying to break down.

The recent resurgence of purpose-type explanation in biology

14. Cited by George Lukacs in the opening paragraph of *The Theory of the Novel*, trans. Anna Bostock (London: Merlin Press, 1971), p. 29. For a fuller account of the way in which philosophy has encouraged certain attitudes to nature, compare John Passmore, "Attitudes to Nature," in *Nature and Conduct*, ed. R. S. Peters, Royal Institute of Philosophy Lectures vol. 8 (London: Macmillan, 1975).

15. Friedrich Nietzsche, *Complete Works*, ed. Oscar Levy, vol. 13, *The Genealogy of Morals*, trans. H. B. Samuel (Edinburgh: Foulis, 1910), 3:25.

16. B. F. Skinner, *Beyond Freedom and Dignity* (New York: Alfred A. Knopf, 1971), p. 213.

has increased, not diminished, Ursula-type fears. For they are teleonomic, not teleological in the old sense. They show promise, indeed, of explaining exactly how we can speak of the parts of an organism as having a "purpose" without concluding that in virtue of its purposefulness an organism is totally different from any conceivable mechanism. Earlier, Darwin had already set us on this path. Darwinism allows us to ask and answer such questions as: "What is the purpose of the elaborate courting dances of the Great Crested Grebe?" But the answer shows in detail how these practices form a close mating link between male and female and just how essential this mating link is for the survival of that particular species. And this is *not* the sense in which human beings have wanted to argue that rituals have a purpose—it does not relate such rituals to the good of humanity.

Whether this fear that science might deanthropomorphize human beings has any grounds is the leading philosophical issue, one might be prepared to say, of our time. Of course, it is not a new question—La Mettrie wrote his *Man a Machine* in 1747. But there was in La Mettrie's time no machine with which human beings could plausibly be compared. Between an inflexibly programmed clockwork and a flexible human being there lies an immense gulf. As science has made it possible to build machines capable of performing to a greater degree what had always been presumed to be uniquely human tasks—like, for example, playing chess—the relation between man and machine has become steadily closer. There have been failures, like the failure to develop a satisfactory translation machine, and a really expert chess-player can still beat a machine. But one should not be surprised that these setbacks and failures are often set aside as only temporary, even if my own conclusion would rather be that they set the limits to routine, the margins of originality. And as science has come to understand the workings of the human body, the body's powers as a mechanical system become daily more striking. It would be *obviously* silly to explain Socrates' remaining in Athens in terms of his muscles; it is far from being *obviously* silly to explain it in terms of his brain, his nervous system, his glandular secretions or, as Skinner does, as a combination of his genetic constitution and his past history, which has rewarded certain of his actions and penalized others.

No doubt, many philosophers have argued that it is absurd to suppose that concepts like "intention," and "responsibility" will ever turn out to be theoretically dispensable, translatable either into Skinner's language or into the language of physiology.[17] But the fear of science often rests on a fear that they might be wrong, a fear not uncommonly expressed with particular force by poets and novelists, as by Dostoevsky in his *Letters from Underground* or, as we have already seen, by Ursula Brangwen in Lawrence's *The Rainbow*.

We may be tempted to dismiss out of hand Ursula Brangwen's question: "Was she herself an impersonal force, like one of these?" How, we are inclined to reply, could *she* be an impersonal force? What would then count as a *personal* force? The deanthropomorphizing of nature, we might continue in the same spirit, consisted in deciding that nature was not, after all, like human beings in possessing purposes and intentions. But how could we possibly conclude that human beings were not like human beings?

This way with dissenters, however, is too short. There was a time at which the contrast "personal-impersonal" had no application; whatever existed was taken to have the characteristics we now think of as being peculiarly personal. That time could come again, except that now we would be supposing that what we had taken to be peculiar to the impersonal was universal in its application. And to say that "she" could not be impersonal is a mere verbal trick; not all languages contain personal pronouns of the English sort. So considerably more substantial arguments would be necessary in order to prove against Skinner that human beings are not the victims of a myth—a myth, Skinner argues, which once had survival value—a myth about their own nature, the mythical belief that they are autonomous beings. Skinner is not denying, of course, that human beings are different from the nonhuman—any more than the chemist denies that chalk is different from cheese when he surprises us by his discovery that they are both compounds of calcium. But he does deny that human actions can only be properly explained in ways which wholly differ from the ways in which we explain nonhuman actions.

17. Compare the discussion of Skinner in John Harvey Wheeler, ed., *Beyond the Punitive Society* (London: Wildwood House, 1973) and especially the essay by Max Black. I have discussed Skinner in John Passmore, *The Perfectibility of Man* (New York: Charles Scribner's Sons, 1970), see especially pp. 168–69, 171–72.

How does the explanation of human actions in scientific terms amount to a "belittling of man," as Nietzsche suggests? Does it destroy man's belief in his own irreplaceableness, does it convert him into "animal—literal, unqualified and unmitigated animal"? There is certainly something paradoxical in the suggestion that it has had, or could have, this effect. For the mere existence of science itself—the scientific enterprise—makes of human beings something irreplaceable, distinguishes them most sharply from any other inhabitant of the earth's surface. Not only in power—this raises the question I shall take up in the next chapter—but also in respect to understanding. Were there no human beings there would be no science. In that respect, and it is very far from being the only one, human beings are irreplaceable. That is why Skinner can write: "Man has not changed because we look at him, talk about him, and analyze him scientifically. His achievements in science, government, religion, art, and literature remain as they have always been, to be admired as one admires a storm at sea or autumn foliage or a mountain peak, quite apart from their origins and untouched by a scientific analysis."[18] The difference is only that we should no longer think of such achievements as involving a peculiar *credit* to their creator, the sort of credit proper only to the creations of a wholly autonomous being.

We do not ordinarily "admire" a view in quite the same way in which we admire a symphony: our admiration for a symphony is, in part, an admiration for its maker. When men believed that God had created the world just as it now stands, their admiration for autumn foliage could be, at the same time, an admiration for its creator. But they have learnt to reconcile themselves to the view that autumn foliage is the product of evolutionary processes, not a deliberate creation. Skinner would have us adopt a similar attitude to what human beings produce.

There is at this point an interesting similarity between Skinner and the powerful Christian tradition that God's grace, not individual effort, should be given the credit for what we normally describe as "our" achievements.[19] Notoriously, however, when, as happened in seventeenth-century New England, Christian theologians

18. Skinner, *Beyond Freedom and Dignity*, p. 213.

19. See my observations on this theory in *The Perfectibility of Man*, pp. 288–89, and Michael Novak's contribution in *Beyond the Punitive Society*, pp. 233–35.

so emphasised divine grace as to reduce individual moral effort to impotence, many of the faithful either gave themselves up to damnation, convinced that God's grace had not been granted them, or inertly waited in the expectation of grace, until it should be vouchsafed them. Are there comparable dangers in Skinner's interpretation of the human situation? Would men go on creating if they took the Skinnerian view of themselves, if, that is, they no longer accepted the admiration for what they produce as, in any measure, a tribute to themselves? Like their Puritan predecessors, are they not, indeed, already tending to excuse themselves from effort by pointing to their heredity and the social conditioning to which they have been subject?

Skinner argues to the contrary. Human beings create, on his view, because their social experiences have been such as to induce them to do so; similar social experiences should be deliberately engineered by psychologist-kings rather than left to chance. But our experience of absolute rulers more strongly suggests that human beings would be engineered out of creativity into conformity. John Locke was one of the first to argue that human beings could be molded into any desired shape by education. And it is worth remembering that he wrote thus of parents who find their child inclined to poetry: "If he have a Poetick Vein, 'tis to me the strangest thing in the World that the Father should desire, or suffer it to be cherished, or improved. Methinks the Parents should labour to have it stifled, and suppressed, as much as may be."[20] Even if it is possible in principle so to arrange matters as to ensure the continued existence of creative persons—and in fact we have not the slightest idea how, let us say, deliberately to educate a human being to be a Shakespeare or a Plato—one would be certainly rash to assume that these are the sort of human beings a scientific manager would produce or would wish to produce.

But let us return to the more fundamental question of whether the growth of science could force us to abandon such concepts as freedom, dignity, responsibility, as having no real application. At this point we need to make a fundamental distinction between a particular metaphysical interpretation of such concepts and the

20. John Locke, *Some Thoughts Concerning Education,* sec. 174, reprinted in *The Educational Writings of John Locke,* ed. J. L. Axtell (Cambridge: University Press, 1968), p. 284.

concepts themselves as we deploy them in everyday life. On the metaphysical interpretation, they all rest on the idea of the person as a wholly autonomous being: a person deserves credit only when he owes absolutely nothing to others; he is free only if he can act in a manner quite independent of his genetic constitution, his history, and his present circumstances; he is responsible only when no one else, no event in his past history, is in any way a contributing cause of his action; he preserves his dignity only when he perceives himself as metaphysically independent of the general course of events. Thus interpreted, I should agree with Skinner: credit, responsibility, dignity, cannot survive serious inquiry. But to say this in no way destroys our everyday distinction between freedom and compulsion, dignity and degradation, responsibility and irresponsibility, what should and should not be credited to us.

However much we learn about Shakespeare's circumstances, however deeply we explore his family, his friends, his education, the sufferings he endured and the pleasures he enjoyed, the literary tradition he inherited and the social circumstances of his time, he and no one else can take the credit for writing *Hamlet*. None of these discoveries about his history and circumstances is at all like discovering that he only put his name to plays actually written by someone else—in which case he would, indeed, deserve no credit. Equally, however profoundly I have been influenced by others, I am responsible for what I am now writing and I am not, I hope, arguing irresponsibly. I am writing freely, not under compulsion. There is a difference between what I write intentionally and what is a mere slip. I share in the dignity of a free citizen; I do not have to endure the degradations of a concentration camp. No doubt science, through technology, can be used to destroy freedom, dignity and responsibility. Huxley's *Brave New World* is a picture of a society in which this has happened as, in some measure, is Skinner's own *Walden Two*. But scientific analysis does not of itself destroy them; indeed, it exemplifies them. And even in utopian or dystopian societies there are those who must take the *responsibility* for destroying freedom and dignity.

Is all this enough? Or is the mythology of the purely autonomous being in some way essential to our liberal, democratic society?

Science is not man's sole peculiarity; more characteristic, much more universal, is his mythology. It has been widely supposed, under positivist influence, that mythology is simply pre-science, an attempt to explain the natural world without the benefit of scientific method. Indeed, in an essay published as late as 1962 Hempel wrote thus: "In times past questions as to the *what* and *why* of the empirical world were often answered by myths; and to some extent, this is so even in our time. But gradually, the myths are displaced by the concepts, hypotheses, and theories developed in the various branches of empirical science, including the natural sciences, psychology, and sociological as well as historical inquiry."[21] A myth, on this view, is simply a false theory; science replaces it by a true theory—an obvious and unmitigated gain. But at least since Malinowski's essay, "Myth in Primitive Psychology," many anthropologists have seen the situation very differently. "Myth," writes Malinowski, "is not an intellectual explanation . . . but a pragmatic charter of primitive faith and moral wisdom." And again, "it safeguards and enforces morality; it vouches for the efficiency of ritual. . . ."[22] Science, very clearly, does none of these things. On the contrary, it *denies* the efficacy of ritual, or at least its efficacy in any but a psychosociological sense. And it has encouraged the development of a spirit of criticism which tends to weaken, rather than strengthen or enforce, traditional morality.

Malinowski, one might argue, somewhat exaggerates; myths *do* in some measure set out to explain, as the myth of the Fall sets out to explain man's failure to live up to his ideals. But one must still grant to him that, more characteristically, they justify rather than explain. So the Genesis story attempts to justify man's domination over nature, his multiplication, the treatment of women as inferior beings. As these examples suggest, however, the destruction of a myth can carry with it a moral advance, not a moral retrogression. "You are the devil's gateway," wrote Tertullian of women, using Genesis in his justification, "*you* are the unsealer of that (forbidden) tree; you are the first deserter of the divine law; *you* are she who persuaded him whom the devil was not valiant enough to

21. Carl Hempel, "Explanation in Science and in History," in *Frontiers of Science and Philosophy*, ed. R. G. Colodny (Pittsburgh: University of Pittsburgh Press, 1962), p. 9.

22. Malinowski's paper was first delivered as an address in honor of Sir James Frazer at the University of Liverpool in 1925. It can conveniently be read in Bronislaw Malinowski, *Magic, Science and Religion* (Glencoe, Illinois: Free Press, 1948), p. 79.

attack. *You* destroyed so easily God's image, man. On account of *your* desert—that is, death—even the son of God had to die."[23] One can easily see why another anthropologist, David Bidney, more than a little at odds with Malinowski, should say of myth that it "originates wherever thought and imagination are employed uncritically or deliberately used to promote social delusion."[24] The most we can say is that the practices the myth justifies—as Exodus justifies the Ten Commandments—are not *necessarily* undesirable.

The scientific humanist ideal, one might say, is a world without myths, in which men will see for themselves, without feeling the need for any mythical justification, that to be at once rational, free, and loving is the only life proper to man. It is a splendid ideal, one that I do not wish for a moment to deny. But when we look at other societies, we see clearly enough the way in which their moral policies are supported by myths—the myth of the proletariat, the myth of the classless society, for example. And we know that to the inhabitants of such societies, these doctrines are not myths—modern versions of ancient myths about the rule of the saints in an earthly paradise—but manifest truths. Is humanist-democratic thinking dependent on similar myths, myths about responsibility, freedom, and creativity that the growth of science might destroy? Skinner clearly thinks so. I have taken the opposite view. I have argued that science can destroy myths about the degree of autonomy human beings possess without weakening our everyday concepts of freedom, responsibility, and dignity. But if I am wrong, then science could end by destroying the self-conception on which its very existence depends. And that I am wrong, many antiscientists would argue.

Whatever our conclusions on this point, there is certainly one question relating to man that science sets quite aside. "Let any Cause-and-Effect Philosopher explain," as the antiscientist Carlyle puts it in his *Sartor Resartus*, "not why I wear such and such a Garment, obey such and such a Law; but even why *I* am *here*, to wear and obey anything."[25] The question "Why I am here?" in its

23. Tertullian, "On Female Dress," book 1, chap. 1 in *The Writings of Tertullian*, 1:304–5 (Anti-Nicene Christian Library, vol. 11).

24. David Bidney, "Myth, Symbolism and Truth," in *Myth: A Symposium*, ed. T. A. Sebeok (Pittsburgh: University of Pittsburgh Press, 1954), pp. 13, 14.

25. Thomas Carlyle, *Sartor Resartus* (London: Chapman and Hall, 1901), book 1, chap. 5, p. 28.

classical sense undergoes the same fate as the question of why there is anything at all; it is no longer thought of as setting a problem. No doubt the scientist can answer it easily enough, if it be interpreted in a certain way: you are here because your father's sperm fertilized your mother's ovum. But that is not what, classically, the question meant; it meant "What purpose does my existence fill?" And not again in a sense that admits of such a consoling answer as "You support your wife and children, you sell insurance for the support of others." Once more a certain sort of intellectual and moral asceticism is required to set this question aside, except in so far as it admits of such answers as these. Yet the whole thrust of Darwinism is that this is an improper question—that, in fact, "we're here because we're here."

We have so far been discussing metaphysical qualms, deep-rooted fears about what science might become, dissatisfaction because it seems to intensify, rather than to reduce, familiar forms of unease. This already helps us to understand why occultism is now experiencing so conspicuous a revival. But fully to understand this revival, we shall also need to take into account rather different objections to science—objections which still relate, however, to its supposed theoretical inadequacies.

Science leaves a great deal unexplained; it allows, as the sorcerer, the occultist, does not, that there are accidents, coincidences. This follows, indeed, from the very nature of scientific explanation. For such explanations depend on the existence of relatively isolated systems. One can explain, in principle, even if only in principle, why a piece of masonry falls from a wall at a particular time—in terms of weathering, traffic shocks, atmospheric pollution and the like; one could explain, in principle, why, when the masonry fell, a cat was crossing the square. But it remains a coincidence that the piece of masonry hit the cat, in the sense that it is by mere chance that these moments in distinct causal systems coincided. (I am assuming that it is indeed a coincidence, that the cat did not scream at a pitch which brought the masonry down or that it was not enticed across the square when it was known that masonry was about to fall. What we suppose to be a coincidence may, in a particular case, turn out not to be a coincidence. But to say as much is not to deny but rather to rely upon the distinction.)

If in order to explain the behavior of the masonry, we had to take into account the movements of everything it might hit when it fell, explanation would be totally impossible. Neither could we explain the cat's movements if we had to take into account the weathering of the masonry. This does not mean that changes occur without cause. We can explain, in principle, why the masonry fell at a certain time, we can explain, in principle, why the cat moved across the square at a certain time; what we cannot explain, what we do not regard as a candidate for explanation, is why these two times were the same. That we call a "coincidence." Similarly it is a matter of sheer chance that John Smith rather than some other ticket holder won the lottery or that the same number won a prize in two consecutive lotteries. When, as happened recently in Australia, a single ticket, returned each time to the casket, won three prizes in a lottery although the odds against that happening were twenty-five million times a million to one, we were astonished. In this case the odds are readily calculable. But what would be the odds against my mentioning these facts to a particular audience in a particular place on a particular time? I could utter just these sentences, in the first place, only because the long odds in the lottery came good. But even apart from that, one would have to multiply those astronomical odds by millions more to cover the possibilities. Yet lecturer and audience came together, and not as a result of any of those mysterious contra- or suprascientific acausal forces which some would wish to invoke in order to explain the occurrence of the improbable.[26]

"Many empirical concepts," Kant once wrote, "are employed without question from anyone. . . . But there are also usurpatory concepts, such as *fortune*, *fate*, which, although allowed to circulate by almost universal indulgence, are yet from time to time challenged by the question: *quid juris* . . . no clear legal title, sufficient to justify their employment, being obtainable either from experience or from reason."[27] They have a clear legal title, I have been suggesting, if they are used as metaphors, to convey that we are sometimes the gainers from, and sometimes the losers by, coincidence. But not if they are thought of as a kind of cause, or at least an explanation.

If it is not a matter of chance that John Smith won the lottery,

26. See Arthur Koestler, *The Roots of Coincidence* (London: Hutchinson, 1972), part 3, sec. 1.

27. Immanuel Kant, *Critique of Pure Reason*, trans. N. K. Smith (London: Macmillan, 1950), B117, A85.

then the lottery was improperly conducted. This the occultist will not admit. There must, as he sees it, be some reason why John Smith rather than Mary Brown wins the lottery—some reason other than that the one has bought, and the other has not, a ticket with the winning number. Such beliefs are, I think, still quite widespread; a notable coincidence causes a curious kind of excitement, perhaps a shiver, even although we all know, if we think about the matter at all, that coincidences occur every moment of the day. We notice with astonishment the lottery ticket which bears our telephone number; it would be no less, and no more, a remarkable coincidence, if it bore a number precisely two thirds of our telephone number—a fact we would not even notice. The winning number is bound to have some unique relation to our telephone number, whatever it is.

The recognition of coincidences as being simply that, not a necessity in disguise, takes away from us a degree of consolation in our sorrows. It would have been a consolation for Job to know that his afflictions formed part of a cosmic struggle. It was a consolation, even, to feel that there was someone he could hold responsible for them. His afflictions, thus envisaged, had a certain point—just as to die *for* a cause is more satisfactory than to die *from* a cause. Indeed the belief that we die simply *from* a cause is one of the hardest of all doctrines to accept, the most stoical, in the sense in which I have been using the word.

The hostility that a great many people feel for euthanasia does not derive from the quite rational fear that euthanasia might be, as it has been, misused as a political weapon. To permit euthanasia would be to grant once and for all that the time and manner of a man's death is not "part of a plan." But that, I should say, is precisely what the scientific analysis of disease and death has brought home to us. Medical science explains our death, but not in any sense which gives to our death a cosmic significance. And social science, too, explains what happens in ways that do not permit us to *blame* anybody. As soon becomes apparent in times of crisis, the desire to find some particular individual, or group of individuals, whom we can blame when anything goes wrong is still deeply rooted in our ways of thinking.

At this point, dissatisfaction with scientific explanation closely relates to dissatisfaction with scientific prediction. Just because so much of what interests us is a matter of chance, we cannot predict it with certainty. We cannot predict with certainty how long John Smith will live, because he may tomorrow be struck on the head by a piece of falling masonry. If a modern Macbeth sought to know whether his children would rule after him, he would still have to seek an answer from witches, not from scientists. Of course, the scientist can predict a great deal which was previously unpredictable—that, for example, a space ship will reach the moon. If that prediction does not come off, it is as a result of a technical fault which further investigation can remedy. But such cases are not typical; they depend on the fact that the solar system is, relatively speaking, a closed system. Most scientific prediction, outside such exceptional closed worlds, is statistical in form; we know what the chances are that John Smith, considered as a member of certain reference-classes, will live to be seventy. But this is as far as we can go by the very nature of the case. It is hard for people to reconcile themselves to this kind of limitation—"a rough rule of thumb," writes Roszak, "the more statistical the mode of discourse, the less its personal relevance"[28]—and not surprising that they turn to palm-readers or astrologers, with their absolute, individualized, predictions.

To return to explanation. Once again, I have suggested, science sets a question aside as unanswerable in principle: the type of question typified by "Why did John Smith win the lottery?" And, as Jacques Monod has argued with particular force, it turns out that the question "Why are men here at all, on the face of the earth?" belongs to this same class.[29] It is a matter of chance that men exist at all; their existence depends on chance mutations, the fact, let us say, that a burst of cosmic rays struck a particular cell. We do not have to think very hard, indeed, to realize what a matter of chance it is that we ourselves, as precisely the individuals we are, exist.

I hope I have now said enough to suggest why it is that antiscientists—or scientists themselves in their extrascientific speculations—are often so profoundly dissatisfied with scientific explanations. Such explanations simply do not answer the ques-

28. Roszak, *Sources*, p. 3.

29. Jacques Monod, *Chance and Necessity*, trans. A. Wainhouse (New York: Alfred A. Knopf, 1971), pp. 138ff.

tions they want to have answered: "Why does anything exist at all?" "What purpose does my existence fulfil?" "What is the ultimate justification of my accepting my moral principles?" "Why did that happen to me?" Or, at least they do not answer them in a way that satisfies the intent which lies behind the questions.

I have argued, in reply, that insofar as antiscience rests on the view that science does not explain at all, that it does not discover "causes and secret motions," it rests either on an exceedingly narrow conception of science or an impossibly demanding concept of explanation; insofar as it rests on the view that science does not answer the really vital requests for explanation, it condemns science for not doing what by the very nature of the case *cannot be done*. But there is a certain slickness in ending thus. Few, if any of us wholly and consistently succeed in freeing ourselves from the kind of thoughts which we officially proscribe as superstitious. I do not accept that fact as a refutation of what I have said. But I should be lacking candor if I failed to make the admission.

Chapter 2
Antiscience and Antitechnology

Let us turn now to the second of Bacon's aspirations: that human beings should use their scientific knowledge in order to extend their dominion over nature. Until the 1960s it was generally presumed, if not without protest from a Blake or a Dostoevsky or a Valéry, that Bacon's ambitions were well on the way to fulfilment. Perhaps, even, that is still the most common view. But a sizable body of dissentients would interpret our situation in quite the contrary sense. Science, they would say, has through technology increased rather than diminished our helplessness, unleashing forces we cannot control. So, according to Jacques Barzun, "the technology of automation, nerve gas, germ warfare and atomic destruction have driven home the lesson of human helplessness directly proportional to human 'control over nature'."[1] Whereas for Bacon, that is, human helplessness and technological progress were *inversely* proportional, for Barzun they are *directly* proportional. And Theodore Roszak has spelled out this helplessness in more detail. Science, he says, has brought in its train technological elitism, affluent alienation, environmental blight, and nuclear suicide.

I do not propose to dispute the view that our world is now, as it generally has been, in a pretty mess. The only point at issue is whether we can properly blame science for our present misfor-

1. Jacques Barzun, *Science: The Glorious Entertainment* (London: Secker and Warburg, 1964), p. 5.

tunes, either in itself or in virtue of the technology it has made possible. Or does the fault lie quite elsewhere?

As so often, the official Marxist answer is attractively straightforward. Science is in no way to blame; neither is technology. The fault lies with the capitalist system, which cannot contain the productive forces it has created. That, indeed, is now the principal gravamen of eastern European attacks on Western capitalism. The "increasing misery" Marx predicted, has, on their view, indeed come to pass—not to be sure, as material poverty, but as a sense of alienation, of impotence in the face of an unmanageable world.

Looking more closely, in the light of these polemics, at Roszak's list of horrors, we can certainly concede to the Soviet critics that "affluent alienation" is peculiar to liberal-democratic societies. But only because so is affluence. Technological elitism, on the other hand, is certainly not a peculiarly liberal-democratic phenomenon. In such Russian science films as Khrabovitske's *The Taming of the Fire* it is quite open and unashamed. And even Soviet apologists will admit to a degree of anxiety about "environmental blight" and "nuclear suicide."[2] Neither do they silently pass over, to take another issue which has aroused concern in democratic countries, the dangers inherent in "genetic engineering." "The danger of ignorant and irresponsible intervention into mankind's genetic information," they indeed tell us, "increases along with progress made in new methods of genetic engineering."[3]

These considerations are fundamentally important, not just a *tu quoque* against the Soviet Union. For they suggest that although the character and the direction of technological innovation will obviously vary in degree and direction from one country to another—in relation to its wealth, its method of capital formation, its emphasis on consumer goods or weapons of war, the strength or weakness of its liberal tradition—the consequences of technology are to a surprising degree independent of such factors. They flow from the nature of technological advance rather than from the political structure of the society. When the Marxist, Ernst Fischer, describes our present fears, he does not pretend that they are peculiar to the democracies:

2. USSR and Czech Academies of Sciences and the Institutes of Philosophy, *Man, Science, Technology* (Moscow and Prague: Academia Prague, 1973), pp. 224–25.
3. Ibid., p. 70.

> There are moments when technical achievements—the flight into the cosmos, which is the realization of an ancient, magic dream—can enchant men. But it is precisely this same power over the forces of nature that also intensifies a sense of powerlessness and arouses apocalyptic fears. . . . A single misreading of a radar report, a mistake by a simple technician may mean world disaster. Humanity may be destroyed and no one will have wanted it to happen.[4]

My central concern, however, is not with technology but with science. Even granted, it might be argued, that technology, both in capitalist and in communist countries, has turned out to be somewhat less of a cornucopia and somewhat more of a Pandora's box than Bacon had anticipated, this is in no way the fault, or even the responsibility, of the scientist. The scientist's heart is pure, his intentions above reproach; he cannot be blamed if the greedy, the lazy, the vain, and the arrogant, seize upon his discoveries for their ignoble ends, whether in pursuit of individual gain or party power. For technology is neither his creation nor under his control.

This is a question—the responsibility of science for technology—on which the scientific community has changed its mind more than once during our century. In the opening decades, the tendency was for scientists to glory in the practical uselessness of what they were doing. Such, according to Duhem, was the tradition in which he was trained in France. At that great center of English science, the Cavendish Laboratory at Cambridge, the same attitude prevailed. "We prided ourselves," so C. P. Snow has told us, "that the science we were doing could not, in any conceivable circumstances, have any practical use."[5] From Cambridge came Bertrand Russell's essay in praise of "useless knowledge." If science—agriculture apart—was in this period ignored by the American government, this was because scientists were taken at their word. There was, so it appeared, no better reason for extending public patronage to the sciences than there was for extending it to the arts. And in the American context that meant—no reason at all.

In the immediate postwar world the situation was very different.

4. Ernst Fischer, *The Necessity of Art: A Marxist Approach,* trans. Anna Bostock (Harmondsworth and Baltimore: Penguin Books, 1963), p. 86.

5. C. P. Snow, *The Two Cultures: And a Second Look* (1963; reprint ed., Cambridge: University Press, 1965), p. 32.

Cambridge's "useless" investigations into the structure of the atom had turned out to have the kind of importance which governments are only too ready to recognize. Scientists took advantage of such facts to obtain governmental support for pure science. In the United States, a National Academy of Science report represented pure science, in language which business-minded legislators could understand, as an "overhead" which had to be met if science was "to pay a dividend" in the form of technological achievements.[6] Scientists, even philosophers, sought funds from military and naval agencies for work which, to put the point mildly, was unlikely ever to contract an enemy's heart with fear. Poets and musicians would perhaps have had stronger claims, relying on the examples of "The Red Flag" or "The Marseillaise."

More recently, confronting a growing reaction against technology, scientists have often tried to have it both ways, at once dissociating themselves—as no responsibility of theirs—from the sort of technology its critics now find abhorrent and claiming credit for the technological achievements which are still generally admired. If, as scientists so often complain, the general public is hopelessly confused about the relationship between pure science and technology the fault is by no means entirely the public's; the propaganda of the scientists has bemused them.

How exactly, we certainly need to ask in the light of this history, does science, as distinct from inventive engineering, offer men domination over nature? Should science more properly be regarded as, to quote the title of Barzun's book, a "glorious entertainment"? Merely as science, one must begin by insisting, it certainly offers us no such dominion. That is why modern science, to say nothing of its Greek ancestor, could flourish for two centuries or so before inventors could make any substantial use of it. Considering, even, the science of our century, the knowledge that organic molecules exist in interstellar space or that the oldest moon rocks are of such and such an age does not by *itself* offer us any greater control over nature. Rutherford was wrong when he asserted, to the very day of

6. Compare S. E. Toulmin, "Intellectual Values and the Future," in *Knowledge Among Men*, ed. P. H. Oehser (New York: Simon and Schuster for the Smithsonian Institute, 1966). On prewar American science see Don Price, "The Established Dissenters," in *Science and Culture*, ed. Gerald Holton, Daedalus Library vol. 4 (Boston: Houghton & Mifflin, 1965), pp. 116–18.

his death, that his investigations into the structure of the atom were, and would remain, technologically useless. But he could have been right; it did not follow automatically from the theoretical importance of his discoveries, as is now so often presumed, that they were bound to be technologically applicable. Perhaps we shall never learn to harness the power of nuclear fusion.

In certain respects, indeed, science by its very achievements can increase, if not our actual helplessness, then at least our feeling of helplessness. As Sir Karl Popper has particularly emphasized, it lays down the limits within which human ingenuity has to work, tells us, for example, that we cannot dump uranium waste in the expectation that it will rapidly cease to be dangerous to life and health, cannot construct a perpetual motion machine, cannot hope to change the weather in one part of the world without changing it in any other part or to contain nuclear fusion within a conventional engineering framework. That is one cause of resentment against science. It often says: "that is impossible." A witch doctor or a "spiritual healer" never tells his patients that there is absolutely nothing to be done; magical powers know no limits except more powerful countermagic.

In other respects, too, science takes away from human beings powers which they confidently believed themselves to possess. "It would be better to follow the myths about the gods," Epicurus long ago wrote, "than to become a slave to the physicist's destiny. Myths tell us that we can hope to soften the gods' hearts by worshipping them, whereas destiny involves an implacable necessity."[7] And the phrase Epicurus here employs—"the physicist's destiny"—reminds us that Whitehead once derived modern science from ancient tragedy. "Their vision of fate, remorseless and indifferent, is the vision possessed by science. Fate in Greek tragedy becomes the order of nature in modern thought."[8]

Science, then, incorporates the tragic view of life, for all the unrelenting optimism which some of its proponents have exhibited. In Dostoevsky's *Letters from Underground*, still the most striking of all antiscientific literary works, his narrator describes

7. Epicurus, "To Menoeceus" in *Epicurus, The Extant Remains*, trans. Cyril Bailey (Oxford: Clarendon Press, 1926), par. 134 (Adapted by the author).

8. A. N. Whitehead, *Science and the Modern World* (Cambridge: University Press, 1926), p. 14.

scientific laws as "stone walls," which scientists are quite prepared to accept as such—as, indeed, in their very finality calming, even mystical—but which the narrator sees as an intolerable limitation on his freedom. Science compels us to face the fact—as archaic Greek religion also taught but Christianity, uncomfortable with predestination, only ambiguously—that there are laws which have simply to be recognized, which are not modifiable by prayer and entreaty, by a plea for mercy, or by the excuse, however justified, that we were not to blame since we acted out of ignorance. Nothing has brought that fact more forcibly home to us than the discoveries of ecologists.

What extends our dominion over nature is, indeed, not science as such but invention—not Newton's law of gravitation but the parachute or the jet engine. Bacon and Descartes saw this, saw that science could not by itself compete with the pretensions of magic. To do so it needed the help of the artisan, the mechanic. There is some controversy about the precise role which science played in the Industrial Revolution. But for the most part the great inventions of the eighteenth century were still the work of practical men, suggesting, not relying upon, a general theory of power and energy. A developed theory of aerodynamics, similarly, was subsequent upon, not prior to, the success of inventors in constructing effective flying machines by trial and error. And even now, the majority of everyday inventions owe little or nothing to science or depend only on science of the most elementary sort, so old-fashioned as to have been incorporated into common sense. They are the work of "practical men," where "practical" is so defined as to be antithetical to theory, as involving a "know-how" that carries with it the very minimum of "knowledge-that."

It is time, however, for us to make a little more precise a threefold distinction on which we have so far implicitly relied: between pure science, practitioner inventions, and technological inventions. Pure science attempts to discover general principles of the widest possible generality—laws and theories. (I am not suggesting that this is *all* it does.) The degree of generality varies, by the very nature of the case. So biochemistry is less wide-ranging than chemistry. When a science is dependent on the laws of some other science, we sometimes speak of it as an "applied science." I have heard it argued, indeed, that the only pure science is mathematics.

But this is a distinction which is unimportant for our purposes—in spirit and aim, in their criteria of success, biochemistry and geochemistry do not differ substantially from chemistry. Let us only remember that, in my usage, "applied science" does not mean, as it often does, *practically* applied; it signifies, only, a degree of *theoretical* dependence on other sciences.

But if pure and applied sciences, in this sense of "applied," can, for our present purposes, be assimilated, the distinction between practitioner inventions and technological inventions must be insisted upon. Both create devices—machines, procedures, methods of storage—which, when they are successful, help us to attain our ends in what we regard as an economical manner. (The word "economical" has here to be interpreted very broadly—as meaning "in the manner which maximises benefits and mimimises costs.") The difference between the two, as I propose to use these words, is that technological inventions, unlike practitioner inventions, are science-dependent. So the violin is a practitioner invention, the moog synthesizer a technological invention; the mediaeval clock is a practitioner invention, the quartz-crystal watch is a technological invention; the manuring of fields is a practitioner invention, the use of chemical fertilisers a technological invention. For my purposes, and unlike many of the critics of technology, I also want to confine the phrase "technological invention" within boundaries which would exclude what I should rather call "bureaucratic innovations"—of which the metric system is one example, codified systems of law another. These resemble technological inventions in so far as they are introduced on theoretical grounds to improve procedures. (Contrast the old practitioner system of weights and measures with the metric system, or common law with the Napoleonic code.) But although they arise out of some sort of systematic thinking it is not that peculiar kind of thinking which we call "science." Bureaucratic innovators flourished in civilizations like imperial Rome, which were rich in engineers and lawyers but had no scientists. It would be quite wrong, then, to praise or blame *science* for such innovations, even when they are the work of intellectuals.

To return to our main theme, physical science and practitioner invention have been closely allied as long as science has been experimental. At least since the sixteenth century, the scientist has made use of the artisan's skills—the skills of the lens grinder, the glass blower, the instrument maker—to facilitate his observation and understanding of the world; he has advanced invention, in other words, as a consumer. Even the scientist's more general theoretical concepts at first derived from practitioner inventions—as when Harvey thought of the heart as a pump, Boyle of atoms as billiard balls. The problems the scientist faces, furthermore, have often been suggested, as in the case of aerodynamics and the airplane, by practitioner's devices. Only over the last century or so has the physical scientist repaid his debt; many of the experimental instruments now in use could only have been created with the help of scientific principles.

If we blame scientists for, let us say, "environmental blight," they are likely to reply—making use of the distinctions I have just sharpened—that the villain, if there is one, is *invention*, not science. And they are so far justified in that many of the most ecologically destructive innovations—the agriculture and the irrigation which gave rise to the mid-Eastern deserts—were in no sense the work of scientists. But, as the very definition I have offered of it suggests, it is not so easy for scientists to avoid responsibility for *technological* inventions, dependent upon science for their very existence.

True enough, there is usually a gap, often a very considerable gap, between scientific discovery and technological invention. The invention of cathode tubes to make possible the experimental testing of electronic theory did not entail the construction of television sets, any more than Newtonian mechanics entailed spaceship exploration. There must first be a demand for television sets, for spaceships; otherwise they would not be built. But if science does not create the demands which technological invention satisfies it does reveal the possibilities which technology realises. And sometimes, as in the case of spaceships, it creates, or at least strengthens, the demand. In many such instances, of course, not only a demand is needed, but also a great deal of imaginative hard work on the part of that underestimated body of creative human beings—engineers and designers. In other instances, the gap between scientific discovery and practical application is so easily filled that it scarcely exists. This is particularly true of medical research. As the case of penicillin makes plain, a pharmacological discovery does not always immediately assume a marketable form. But when it does, little more may be needed than a large-scale duplication of the laboratory processes by which the discovery was made or a slight modification of such familiar devices as the tablet or the syringe. Under these circumstances, at the very least, scientists can scarcely refuse to accept the responsibility for innovations which turn out to be ill advised. That may be one reason why biologists tend to be much more conscious of the likely practical outcome of their discoveries than are physicists—the relationship between discovery and application is more direct.

The Spanish Inquisition sought to avoid direct responsibility for the burning of heretics by handing them over to the secular arm; to burn them itself, it piously explained, would be wholly inconsistent with its Christian principles. Few of us would allow the Inquisition thus easily to wipe its hands clean of bloodshed; it knew quite well what would happen. Equally, where the technological application of scientific discoveries is clear and obvious—as when a scientist works on nerve gases—he cannot properly claim that such applications are "none of his business," merely on the ground that it is the military forces, not scientists, who use the gases to disable or kill. This is even more obvious when the scientist deliberately offers help to governments, in exchange for funds. If a scientist, or a philosopher, accepts funds from some such body as an office of naval research, then he is cheating if he knows his work will be useless to them and must take some responsibility for the outcome if he knows that it will be useful. He is subject, properly subject, to praise or blame in relation to any innovations which flow from his work.

This still leaves a range of possibilities open, and in a very important way. The scientist may not *know* whether his work will have technological applications; there may be just a chance that it will. He can ease his conscience as a scientist by appealing to the fact he does not *know* how his work will be applied, as a fund applicant on the ground that he does not *know* that his work will be useless. Neither can he be at all confident that if his work does turn out to have applications their total effect will be undesirable rather than desirable. "This has killed a beautiful subject," the Australian atomic physicist, Sir Mark Oliphant, is said to have observed as the bombs dropped on Hiroshima and Nagasaki. The bombs killed, as well, more than a few people. Oppenheimer's reaction was perhaps more typical: "In some crude sense which no vulgarity, no overstatement can quite extinguish, the physicists have known sin, and this is a knowledge which they cannot lose."[9] But we do not yet know what will be the final consequences of the scientist's work, whether, to take the extreme possibilities, the nuclear bomb will help to preserve humanity from still another large-scale holocaust or whether it will wholly destroy life on earth.

So considerable are the uncertainties, that we can understand why some scientists have argued that we just have to put up with the devastation which the growth of science can bring in its train, if only because we are too ignorant to do anything else. "Man does not sufficiently foresee," writes the physiologist Lord Brain, "the consequences of scientific discoveries. It follows that he certainly has not the capacity to decide that some particular line of scientific research ought to be abandoned because of its supposed evil consequences for mankind."[10] On this view scientists should neither be praised nor blamed by pointing to the consequences of their discoveries. For they are acting, all the time, out of complete ignorance.

But we do not usually adopt the principles here taken for granted. We can never be quite sure that we are acting for the best. Sometimes, as Kant and Hegel long ago argued, "the cunning of history" brings it about that the greedy and the envious produce highly desirable social changes whereas the well-intentioned reformer generates catastrophes. The same is true in our personal lives. Trying to help others, we sometimes destroy them. But these reflections, disconcerting though they are, do not justify us in embarking upon actions which, so far as we can foresee, are likely to lead to disastrous results. Where the consequences would be exceptionally disastrous, we ought to modify our plans if there is even a slight risk that such consequences will come about. (Expressions like "slight" and "exceptionally disastrous" lack precision. Human beings being what they are, of all degrees of temperament between the manic and the depressive, they are bound to differ about the degree of risk-taking which, at the margins, is rational.)

9. Cited in John Maddox, *The Doomsday Syndrome* (London: Macmillan & Co., 1972), pp. 9–10.

10. W. Russell Lord Brain, *Science and Man* (London: Faber and Faber, 1966), p. 87.

The antiscientist, indeed, can plausibly reverse Brain's argument. Since we can never fully predict with confidence the outcome of scientific investigations, his argument would then run, we should not embark upon them, ever, even when it looks as if their outcome would be wholly favorable. But the principle which lies behind that conclusion, too, is not a tenable one. It would, indeed, make any sort of action impossible; we can never be *quite* sure that we shall not break our leg getting out of the bath tub. Human beings live, and always will live, under conditions of uncertainty, and their greatest gains often involve them in the greatest risks, as when they love—or inquire into the nature of things.

If we believe that science is a valuable form of human enterprise—whether it is, we still have to consider—then we shall naturally be reluctant to see it curbed or hindered. So far we can sympathize with Lord Brain. And we shall properly be suspicious of "dangerous consequences" arguments, invoked as they so often are by the censor, the obscurantist. But where the science has as its explicit objective—as is true of much combinatory biology—the relieving of human ills, then we can properly comment that the risk is too great, that this is an uneconomical way of achieving these ends. Neither should we allow ourselves to be assuaged by comforting statistical statements about the actual risks unless we have looked very carefully indeed at the grounds for these statistical estimates, especially when the statistics come from scientists who are naturally desirous to carry on with their own line of inquiry and are not accustomed to take into account any but a limited range of consequences. This still leaves us faced, however, with a moral dilemma which is anything but easy to resolve: if the inquiry will advance human knowledge, our commitment to freedom of inquiry can pull us in one direction, our fear of consequences in another. We can certainly try to ensure that everything possible is done to reduce the risk of adverse consequences by, for example, sealing off laboratories. These precautions, however, obviously do not touch upon the possible *political* misuse of biological discoveries; the laboratory is not sealed off from society and cannot be. (The "mad scientist" of popular fiction, from Dr. Frankenstein onwards, has generally been a life-creating biologist, and it is the social consequences of, let us say, cloning which are held up for us as objects of fear in Huxley's *Brave New World*.) Just what risks we are willing to take for the sake of scientific freedom will obviously depend on the value which we attach to that freedom. But we should never forget that the attempt to restrict freedom can produce precisely those political consequences which make us fearful of the outcome of freedom.

Before we pursue these reflections further, let us look at the question whether science—operating through technology—does in fact increase human helplessness to a degree greater than practitioner inventions. (There are primitivists, of course, who would oppose any sort of invention. In many a story of the Golden Age, human beings went astray as soon as they cooked their food and wore clothes. But so absolute a primitivism is not our present concern.) Very obviously, it might plausibly be replied, science has made human beings less helpless than they once were. For a large percentage of the inhabitants of the world, no doubt, life is still "nasty, brutish and short." But the fact remains that human beings have learned in principle, with the help of science, how to control the major diseases; by flicking a switch they have extraordinary powers at their disposal. If these advantages are still not universal, that is certainly not the scientist's fault. In the light of such facts, how can it be denied that science has enlarged our power to control the world around us?

These arguments do not satisfy the antiscientist. Technological inventions make us feel helpless, he will argue, as practitioner inventions did not, because they are comprehensible only to an elite. With a modicum of intelligence, most people can come to understand the working of inventions based on the lever, the wheel, the cog. A steam engine, as compared with a diesel electric, has indeed something *human* about it. It generates metaphors, like "letting off steam"; it has inspired poetry. We can learn, or envisage ourselves learning, how to repair such inventions. We admire the manual dexterity of a watch repairer or a plumber but have no disposition to feel that he possesses magical powers.

Characteristically, however, science-based inventions involve what has come to be called a "black box." We see what feeds into it—let us say a power line—and we may be able to repair its plug; we see, too, what comes out of it and by adjusting knobs or tightening a screw we may be able to improve its performance. But what lies in between—the very heart of the invention—is to most of us a mystery. If it has to be repaired, we call in the aid of an expert;

most often, he takes the black box away, to test it with the aid of other black boxes, the inner workings of which may be as mysterious to him as the original black box was to us. So Roszak is right enough; technological invention, as distinct from practitioner invention, brings in its train a technological elite; even when we are grateful for their attentions, we are at their mercy. (The technologist, incidentally, often feels much in the same way about the higher reaches of science; he uses what he does not really understand; if something goes wrong, he can only refer the question back to the scientists.)[11]

"Science," Malinowski once wrote, "is open to all, a common good of the whole community, magic is occult, taught through mysterious initiations, handed on in a hereditary or at least in very exclusive filiation."[12] But nowadays it is precisely as magic that science appears to a great many citizens, and not only science but technology—as occult, taught through mysterious initiations, handed on in a very exclusive filiation. Its generation of an elite is one reason why we so often find it alleged against science that although it purports to give human beings power over nature what it really does is to give certain human beings power over others. This is an accusation which makes strange bedfellows. C. S. Lewis and Herbert Marcuse are scarcely soul mates, but agree in putting the charge; in science fiction it is omnipresent by implication. And it cannot be set aside as the panic-stricken reaction of humanists confronted by phenomena which the defects of their education have rendered incomprehensible. Brainwashing, electronic bugging, computerized records, are by no means mere figments of the terrified humanist imagination. What distinguishes them from traditional forms of control—by violence or bribery—is that they rest on a degree of science-based expertise.

In relation to them we are all of us potential victims. Courage and honesty are largely unavailing. Neither can we retaliate; we do not know how to, we do not have the resources at our disposal, even if we could bear to exercise the degree of inhumanity involved.

11. On science as a "black box" see Mario Bunge, "Towards a Philosophy of Technology," in *Philosophy and Technology*, ed. C. Mitcham and R. Mackey (New York: Free Press, 1972), p. 64.

12. Bronislaw Malinowski, *Magic, Science and Religion* (Glencoe, Ill.: Free Press, 1948), p. 3.

And what lies in prospect, in the form of genetic engineering, does not encourage optimism, except among those who are firmly convinced that good men, that is, they themselves and their friends, will do the manipulating. It is all very well for the sociologist Kingsley Davis to tell us that as a result of biological engineering, the universe will be "man's at last."[13] Whose in particular? All of us, or a powerful few?

Many advanced technological products display, to make matters worse, a queer sort of quasi rationality, quite divorced from feeling; it is easy to envisage the computer, for example, as a psychopath. Of course, the scientist does not envisage it thus; the computer is for him—as to a growing degree, even, for the scholar—a magnificent, if occasionally exasperating, research tool. But most of us encounter the computer only at the level of the computerisation of what were once personal transactions. And at that level it is harder to influence than the most obdurate bureaucrat. It performs tasks, with increasing ingenuity, which we once supposed to be peculiarly rational, intellectual, beyond any but human powers. Yet we cannot argue with it, make representations to it, unless it has been programmed to take account of them. As for appealing to its fellow-feelings. . . . As long ago as 1912 E.M.Forster wrote a short story "The Machine Stops" in which a computerized society was represented as a human hell, in which the sole political question was: "How much will it cost to re-programme the machine?" and individual differences were wholly ignored as too costly to repay attention.[14] Sometimes that society seems all too close to us.

Then, too, technological innovation has a terrible way of doubling back on us; just as we are congratulating ourselves on our gains they turn out to be losses. "This accursed science," H. G. Wells once wrote, "is the very Devil. It offers you gifts. But directly you take them, it knocks you to pieces in some unexpected way." Even the control of disease has brought starvation in its train; medical science has greatly increased the chance that you and I will suffer a slow and painful death.

13. Kingsley Davis, "Sociological Aspects of Genetic Control," in *Genetics and the Future of Man*, ed. J. D. Roslansky (Amsterdam: North-Holland, 1966), p. 204.

14. E. M. Forster, "The Machine Shops," in *The Collected Tales of E. M. Forster* (New York: Modern Library, 1968).

We granted, no doubt, that practitioner inventions, innocent of science, can also have unpredictably disastrous consequences. But the more developed our capacity for technological inventions, the more widespread and terrifying these consequences turn out to be. This is no accident; unlike practitioner inventions, technological inventions can release into the biosphere substances quite unlike anything with which it has ever had to cope. Chemistry was the first to do this; it created molecular combinations of a sort which the animal body had never previously encountered and to which it could not adjust by the slow processes of evolution. It continues to do so—but physics now contributes the produce of nuclear fission, biology new viruses. Human beings have never before had to contemplate the possibility—and would never have had to do so, if invention had remained at the practitioner level—that they might day by day be destroying the conditions which make possible human life.

Technological invention progresses, furthermore, by sudden leaps, in response to "scientific break-throughs." Practitioner inventions, in contrast, change only gradually, by a process of slow modification. The gap between a precomputer civilization and a computerized civilization is a tremendous one. It is as if human beings had moved in a few years, rather than over millenia, from tentatively cultivating a few crops to large-scale agriculture, or from the stone axe to the steam hammer. In less than half a century, again, we have moved from a situation in which only a few scientists so much as contemplated the construction of nuclear bombs to a time when we quite rationally fear that they might be built by terrorists. In consequence, the psychological and social effects of technological innovations are overwhelming, to say nothing of their biological consequences; they are on top of us, taking over our lives as thoroughly as television has done, before we have any chance of reflectively judging their character or estimating their consequences. "How can man get into harmony with his surroundings," Forster once asked, "when he is constantly altering them?"[15]

As they stand, these considerations are undeniably powerful;

15. E. M. Forster, "Art for Art's Sake," in E. M. Forster, *Two Cheers for Democracy* (London: Arnold, 1951), p. 100.

they sufficiently explain the feeling that our so-called "control over nature" makes us steadily more helpless, and why Forster should argue that only apathy, uninventiveness, and inertia can save us. No doubt, they will not alarm everybody to the same degree. Greater control of man by man is precisely, so Skinner and the Soviet Marxists would argue, what our society needs; if technology makes it possible, so much the better. Idle talk about privacy, freedom, dignity should not be allowed to stand in its way. Technologists themselves, or so social surveys suggest, are almost to a man intensely conservative, rigorous fundamentalists, happy to work for, and within, a society organized as a machine. "The engineer," as one of them has put it, "is too busy keeping things going to worry about society. That is why he is a natural conservative."[16] "Keeping things going"—even if what is kept going is a gaschamber in a concentration camp. What ought to be kept going is now precisely the point at issue.

Many defenders of science rest their hope of avoiding technological disasters on relatively minor social changes. A better education along with more effective, but still democratic, social controls can, on their view, free science from its dangers. With an improved scientific education, we shall all come to understand the technological innovations which frighten us by their incomprehensibility, to understand them at least in principle. We shall learn simply to shrug off the vagaries of computers, once we understand the principles on which they work—see why we must occasionally send them a check for no dollars and no cents, accept as a minor inconvenience a flood of the same material from the same source. A more broadly based humanistic education for scientists, it is added, will make of them more socially responsible beings, enable them to understand the consequences of their acts. Laws against centralizing computer data or against the invasion of privacy, "environmental impact statements" which take account of social as well as ecological consequences—none of these is incompatible either with the continued development of science or with the maintenance of democracy.

But such reforms, such examples of what Sir Karl Popper, in an

16. W. H. G. Armytage, "The Rise of the Technocratic Class," in *Meaning and Control*, ed. D. O. Edge and J. N. Wolfe (London: Tavistock Publications, 1973), p. 85.

interestingly technological phrase, called "social engineering," still leave the antiscientist deeply dissatisfied. They would dissatisfy, too, the proscience Marxist for some, if only some, of the same reasons. Capitalist governments, capitalist industry, he would argue, are too powerful to permit such efforts at social control ever to be effective. It is one thing to produce an environmental impact statement, quite another thing actively to control "environmental blight." Excuses will be found, evidence will be falsified, bribery and less blatant forms of social pressure will be exercised, to ensure that profitable technological innovations are marketed, however destructive their environmental and social impact. As for the state, and here the Soviet Marxist falls suddenly silent, it cannot be expected to suppress investigations which, if successful, would increase its own power. What is more likely to happen is that, with the help of secrecy, the state will ensure that it has a monopoly over the resulting devices. So control over electronic bugging devices, for example, will have the practical effect that the state can bug everybody, whereas no one can bug the state.

It would be quite foolish to dismiss these criticisms out of hand, whether as political propaganda or an exercise in cynicism. But we can properly resist them with at least a qualified degree of optimism. We can draw attention to the reawakening in recent years of community pressures, which, if sometimes defeated, at least *attempt,* and not always unsuccessfully, to control technological innovations. (Their successes correspond in an interesting way to the successes of guerilla fighters, for long supposed to be engaged in a hopeless task against scientifically based armies.) We can at least be confident that our chance of exercising such restraints is considerably better in a liberal-democratic than in a centralized state. Yet we cannot repress a fear that our very attempt to exercise controls may increase the power of the technologically minded arms of the state—above all, of the police forces. One is not at all surprised to find it argued that the growth of technology, especially if that term is interpreted widely enough to include bureaucratic innovation, is quite irresistible. Whichever way we turn, it might be argued, all that happens is that we replace one form of control by another.

Take the supposition that we can reduce the bad effects of technology by modifying our education system, humanizing our scientists, and bringing humanists to a better understanding of science. Who now runs our education systems? Certainly not humanists. They are run, indeed, by men claiming to be experts, applying psychology, economics, sociology, administrative theory. To expect an improvement in *understanding* from institutions which themselves are dedicated to turning out expertly trained experts is to display, critics like Jacques Ellul would argue, a monstrously limited conception of technology's range and power. (Appealing against technology, on this view, we inevitably appeal from Caesar to Caesar.)

Indeed, technological determinism threatens to take the place of the once influential geographical determinism—not surprisingly, seeing that technology has largely replaced nature as it once was by what Cicero called "a second nature," an environment which is man-made, man-controlled, so that even the climate in which we daily live and work is not nature's climate but one of our own making. The comparison suggests, however, that writers like Ellul may be exaggerating, just as the geographical determinist exaggerated. Technological innovations, it must be granted, profoundly influence our lives; a television set, a car, a computer is not just a tool in the sense that a hammer is a tool, something we can take out of a tool box when we have need of it and then put away again. Once we become involved with such inventions, we start behaving differently—and over a wide range of our activities. The mere possession of a car alters our way of life. And this is true of a society as well as individuals. Once a society has an industry devoted to making automobiles, that industry, by its costs and complexity, makes fundamental social demands on the society which contains it: it limits the society's options. But not quite irresistibly: social patterns do change, the social importance of an invention can rise and fall in degree; individuals can resist its power. The technological determinists should certainly disquiet us: I do not for a moment wish to deny that technology impels us, in certain respects, in the direction of a centralized, bureaucratised, society. The best I can hope for is that this pressure is not equivalent, any more than it is in the geographical case, to complete coercion. Although our technological environment, like our natural environment, limits what we can do, how we can live, it does not wholly determine our actions. But the degree to which it does restrain us must not be underestimated, or the dangers inherent both in the kind of restraint it exercises and in the possibilities it opens up.

Furthermore, or so it is often argued, the technologist's attitude of mind, not only his theories, derive from science—from the metaphysics science presumes, the methods it employs, the ethics it encourages. With this metaphysics, these methods, such an ethics, technology cannot but be devastating in its impact; the banning of DDT, the cleaning up of this or that river, the reduction of lead emissions from exhausts are as laughable, or as tragic, as an attempt to cure oneself of a dangerous disease by going to bed earlier. We may feel slightly better, if only because we are doing something positive. But the disease continues on its way. Nothing short of the ruthless excision of science from our culture can possibly save us.

On this view, science is responsible for our ills, not just because this or that technological application of scientific endeavor has been disastrous, but because it encourages us to regard nature, human beings, our social arrangements, in quite the wrong way. To quote the dissident economist, E. J. Mishan:

> Like some ponderous multi-purpose robot that is powered by its own insatiable curiosity, science lurches onward irresistibly, its myriad feelers peeling away the flesh of nature, probing ever deeper beneath the surface of things, forcing entry into every sanctuary, moving a transmuted humanity forward to the day when every throb in the universe has been charted, every manifestation of life dissected to the *nth* particle, and nothing more remains to be discovered—except, perhaps, the road back.[17]

To suggest that what I called "misunderstandings of science" could be reduced by a better scientific education, such writers would conclude, is not merely utopian but quite preposterous; such a "better scientific education" would in fact be propaganda for science. Whereas Marx once said that only "grammar and the physical sciences, truths that were independent of all party prejudices and admitted of one interpretation, were fit subjects for schools,"[18] what we really need, so the antiscientist argues, is a

17. Edward J. Mishan, *The Costs of Economic Growth* (London: Staples Press, 1967), p. 144.

18. At a meeting of the General Committee of the International, 17 August 1869.

humanistically centred education, from which science would disappear—except perhaps as an example of evil. Grace, not competence, would be its objective.

To attempt entirely to divorce science from technological innovation, to put the blame for our misfortunes on capitalist greed or communist lust for power, is, according to such critics, like trying to absolve Jesus or Augustine of all responsibility for the persecuting spirit of the Christian Church. Their teachings, as the teachings of Buddha did not, lent authority to persecution—"compel them to come in"—even if Jesus and Augustine did not foresee the lengths to which persecution would be carried. Although Rutherford did not anticipate the nuclear bomb, he is none the less, so it is argued, as responsible for its creation as were those scientists who advocated, and aided, its construction. For he *analyzed,* and in the substitution of analysis for contemplation lies the root of all evil. Science, furthermore, is experimental: it is constantly modifying nature to suit its own purposes. Even if these purposes are theoretical, rather than technological, science still sets an example. It encourages the wrong attitude to nature, the attitude which is inherent in technology—as something to be manipulated, not contemplated or enjoyed. And this attitude has spread from inanimate nature to human beings so that people, too, now fall into this same category of manipulable objects. Such are the accusations we have still to meet.

Chapter 3

Antiscience and the Scientific Spirit

Precisely in what respects, we have still to ask, does science, or so it is said, threaten to destroy humanity, not merely in virtue of the technology it makes possible but, more fundamentally, in virtue of the attitudes of mind which it encourages or engenders? Replies to this question come so thick and fast from the critics of science, and cut so deep, as almost to defy summary. But let us make the attempt.

Science, so it is in the first place argued, encourages human beings to think too highly of themselves, of their character, their powers and their prospects. By exaggerating their prowess, it diminishes, in the long run, their actual powers. Secondly, it looks at both man and nature in too abstract, too analytic, a fashion, diverting attention from their complexity, setting aside problems which do not permit of formally neat solutions. Thirdly, just in virtue of this abstractness, with its concomitant emphasis on quantitative mathematical relationships rather than qualitative differences, it threatens individuality, impoverishes the imagination. And all this, finally, in the name of an objectivity which, so far as it is not entirely mythical, is a mask for callousness and political opportunism.

Even although, as our summary already suggests, these far-reaching and formidable objections are intimately linked, they call for separate consideration. That will be our task in the chapters which remain. The first accusation—that science encourages human beings to think too highly of themselves—might well surprise us. For science, so we quoted Nietzsche as asserting, belittles man. "It finds its special pride," to cite him further, "in preserving man's contempt for himself." We had no difficulty in understanding what Nietzsche meant. If to deny, as Kant puts the traditional Christian view, that man is "the titular lord of nature," that he is "born to be its ultimate end," is self-contemptuous, then science certainly encourages self-contempt.

Yet we have only to consider the claims which are still being made for science in eastern European countries—they could easily be matched, until very recently, in western Europe and the United States—to understand why science has been attacked for encouraging us to believe, by its very existence and in defiance of the Greek feeling for *hubris* and Christian humility alike, that a world transformed with its aid, a world which everywhere bears the mark of human intervention, is bound to be in all respects a better world, superior at once in intelligibility, beauty and goodness, to a world which has "just grown." "The Marxist interpretation of the prospects of scientific and technological progress," so a Czech-Soviet statement proclaims, "proceeds from the key assumption that the objective course of history has now led to a situation where all the necessary opportunities and means are available for solving the major problems of human existence, and that this progress is not only a remedy for poverty, famine, disease and hardship but that it turns the human quest for freedom, happiness, personal perfection and creative fulfilment of life into a genuine attainable possibility."[1]

The apparent contradiction between Nietzsche's condemnation of science as "belittling man" and this hymn of praise to its capacity for raising men to the level of gods is readily resolved. Science, in its outcome if not in its intention, at once belittles man's status and aggrandizes his potentialities. For the feudal conception of man as a "titular lord," deriving his power over nature from the exercise of a divinely bestowed right, it substitutes the bourgeois conception of a world which lies wide open to absolute control by the exercise of human talents. And the Marxist questions not that ideal as such, but only the possibility of realizing it within capitalist societies.

1. USSR and Czech Academies of Sciences and the Institutes of Philosophy, *Man, Science, Technology* (Moscow and Prague: Academia Prague, 1973), p. 357.

From this bourgeois point of view, it is not science but Christian theology and Christian-inspired literature which belittles man, by casting doubt on his intellectual and moral perfectibility. Compare, with such a judgment in mind, Shakespeare's Isabella in *Measure for Measure:*

> But man, proud man
> Drest in a little brief authority
> Most ignorant of what he's most assured,
> His glassy essence, like an angry ape,
> Plays such fantastic tricks before high heaven
> As make the angels weep

with Darwin's conclusion to *The Descent of Man:*

> Man with all his noble qualities, with sympathy that still feels for the most debased, with benevolence which extends not only to other men, but to the humblest living creature, with his god-like intellect which has penetrated into the movements and constitution of the solar system still bears in his bodily frame the indelible stamp of his lowly origin.

Who is belittling man and who is aggrandizing him—Darwin or Shakespeare's Isabella? From the standpoint of the traditional theologian, quite certainly Darwin is the belittler. For to speak, as Darwin does, of man's "lowly origin" is to deny that he is made in the image of God. Merely to assert, in Isabella's manner, that he is *like* an angry ape could be, in contrast, an exercise in humility. But if we are thinking in terms of man's fitness to exercise authority over the world then certainly it is Isabella who is the belittler, Darwin the aggrandizer. And in so far as he exalts our potentialities as distinct from our origin, Darwin encourages us, so it is now argued, in our attempt to reshape the world through technology so that it conforms with our ideals, an attempt which a more realistic assessment of our failings—our greed, our vanity, our lust for power—would discourage us from undertaking.

So implausible an optimism, no doubt, is not *peculiar* to science. If Darwin takes too lofty a view of man's sympathy and benevolence, if he ascribes to him a "god-like intellect," this, it might well be argued, is only by way of an *ex parte* statement; Darwin's science, science generally, does nothing to suggest that human beings deserve to be thus elevated. Darwin, on this view, is simply reflecting that optimistic estimate of human potentialities which was first promulgated during the Renaissance, largely dropped out of sight during the Augustinian period which followed, was revived in the seventeenth century and reached its peak in the late nineteenth and early twentieth centuries. It was an optimism derived by combining the Hebraic doctrine that man is made in God's image with a Pelagian denial that Adam's Fall had irretrievably corrupted man's nature. Humanists, philosophers, Pelagian theologians, taught Darwin to write as he did; his conclusions do not in any way flow from his scientific work.

There is a measure of truth in such an interpretation. But science did not simply *reflect* humanist optimism; it seized upon it, gave it a concrete content, lent it plausibility. The chemist-philosopher Joseph Priestley is not atypical. So long as he confined his attention to the political and social history of mankind, he found it hard, Priestley unwillingly confessed, to take any but a gloomy view about the prospects before humanity. But to contemplate the history of science was at once to restore his faith in human perfectibility; man's capacity for social and intellectual progress, he was convinced, shone through that history. As George Sarton was to put the same point a century and a half later, "the history of science is the only history which can illustrate the progress of mankind." Participation in the scientific enterprise generates, of itself, a confidence in human potentialities, contrasting, very strikingly, with the emphasis on human limitations, human failure, so characteristic of the greatest literature, the comic as much as the tragic.

The scientist, furthermore, spends much of his life in a laboratory, surrounded by men very like himself. Imperfect, by no means free of malice, envy, egoism, but engaged, all the same, on common tasks and bringing them, sometimes at least, to a successful conclusion. This experience naturally encourages the supposition that the world, social or natural, ought to be, and could be, managed in the same sort of way. Through his study of literature or history, the humanist, in contrast, is daily confronted by the record of human folly, crime, and misery. Lord Acton's "All power tends to corrupt, and absolute power corrupts absolutely" is the pronouncement of a humanist. And nobody who holds that view will be inclined to believe that power over nature or society is inherently and inevitably beneficial.

The contrast between "shallow" scientific optimism and "deep" literary pessimism—exceptions, of course, abound—has now, as C. P. Snow complains, become something of a platitude.[2] He brings two criticisms to bear on it. The first is that it is by no means obvious, even if the contrast is justified, that the advantage lies with the humanist. Quite justly, Snow draws attention to the way in which some of the best-known twentieth-century poets—Yeats, Pound, and even Eliot—were attracted by a Fascist style authoritarianism, with its pessimistic presupposition that men are not fit to be free, to exercise independent judgment. He should have added, however, that the scientific Cambridge which bred him was no less strongly attracted by Soviet authoritarianism, with its pretence to rest on scientific foundations, its faith in progress. It is surely not a good thing for us to suppose, in Isabella's manner, that whenever we attempt to exercise any sort of authority we are no better than "angry apes," let alone for us to think of ourselves, in the manner of Dean Swift's classical misanthropic pronouncement, as "the most pernicious race of little odious vermin that nature ever suffered to crawl upon the face of the earth." For we may then begin to treat one another as if that is all we were. But it is no less disastrous to suppose, in the manner of the Czech-Soviet statement, that with the aid of science and technology human beings can perfect themselves. That leads in practice to the murder, exile or incarceration, precisely as vermin, of those who "stand in the way of progress." If human beings are not vermin, neither are they gods-in-the-making; the extent of our scientific progress must not be allowed to persuade us otherwise—and has tended to do so.

Snow's second argument is that the description of scientists as optimists rests on a misunderstanding. The scientist, he tells us, can be as depressed as any poet, any theologian, about the more desolate aspects of the human condition. (Some of his scientific critics, interestingly enough, have alleged that in admitting as much Snow spoke for himself as a novelist, not as a scientist.) The scientist's optimism, Snow argues, applies only to the *social* possibilities which lie before humanity. Scientists, he explains, "are inclined to be impatient to see if something can be done: and inclined to think that it can be done unless it is proved otherwise.

2. C. P. Snow, *The Two Cultures: And a Second Look* (1963; reprint ed., Cambridge: University Press, 1965), pp. 12–14.

That is their real optimism and it is an optimism the rest of us badly need."[3] But whether we do "really need" such impatience, such optimism, is precisely the point at issue. Have we not, again and again, moved far too impetuously to remedy specific evils without pausing to consider the broader implications of what we were doing?

Let us look more closely, in the first place, at the view that, unless proof to the contrary is offered, "something can be done." This is certainly an attitude of mind which science encourages. The scientist confronts problems with the assumption that they can be solved by scientific investigation in a manner which will satisfy his fellow scientists; that assumption, indeed, is what keeps him going. He is naturally tempted to presume that what is true of scientific problems must also be true of social problems, that they, too, can be solved, and can be seen to be solved, by a sufficient expenditure of intellectual effort.

A peculiar feature of scientific problems, however, is that it is very clear what counts as a solution. And another peculiar feature is that the question of costs arises, for the most part, only in relation to the expense of conducting the investigation—expenses which, no good scientist will doubt about his own work, ought certainly to be met. In society at large, the situation is very different. The concept of a "problem" and its "solution" is by no means so straightforward. We speak of a "social problem" when there is some phenomenon which we should like to see disappear or, at least, reduced in its incidence—this is the sense in which road-deaths, or crime, or drug taking count as "social problems." The question is not, as it is in science, how some phenomenon is to be *understood* but rather how it is to be *treated*. And there is nothing in the theoretical situation to determine whether a particular phenomenon—rising divorce rates, homosexuality, the decline in religious belief—does or does not constitute a "problem." Equally, there is no theoretical method of determining whether a proposed solution properly counts as such. Road-deaths could certainly be reduced by forbidding the use of the roads between dusk and dawn or by not issuing licences to anyone under the age of thirty. But few would count these proposals as "solutions" any more than they

3. Ibid., p. 14.

would regard large-scale invasions of privacy as a "solution" to rising crime. This is not because they would be ineffective—something which could in principle be determined by a theoretical analysis of the situation—but because the costs, in their judgment, would be too great.[4]

To make the situation more complicated, while everyone agrees that it counts as a "cost" when a scientific experiment involves the expenditure of funds on new equipment, there may be no such agreement on whether it is a cost or a benefit that a technological innovation makes it possible, among other things, to communicate with any person at any place at any time. The costs and benefits, furthermore, may not be commensurable. How are we to "balance" loss of privacy against greater accessibility?

Insofar, then, as scientists are led by the example of physical science to suppose that social problems, like scientific problems, must, in principle, have clear and indisputable solutions, they go astray. Snow's attempted defence of "scientific optimism" draws attention, rather, to the precise reason why it can be so disastrous in its social consequences. In their haste to "do something" scientists and technologists—technologists tend even more strongly to optimism, even if they have a wider appreciation of costs—have until recently ignored the fact that expeditious solutions of technological problems can exacerbate, or even create, social problems.

This attitude is by no means dead and by no means a mere accident. The very structure of science, indeed, has discouraged the scientist from taking up such questions. Social costs—intellectually messy, scattered, obscure, controversial, impossible, or very difficult, to quantify—are not the sort of thing that modern science is geared to take account of. In recent years, admittedly, a number of scientists have come to be perturbed by such costs. But they did not *anticipate* them; physicists working on the atom bomb paid little attention to long-range biological consequences, chemists working on insecticides paid little attention to their effects on biological life cycles. No doubt, as I have already argued, we are never in a position to anticipate *all* the consequences of our actions—indeed, the consequences do not constitute a limited

4. Compare, on these points, John Passmore, *Man's Responsibility for Nature* (New York: Charles Scribner's Sons, 1974), chap. 3.

totality—but in the cases cited above the need for consulting the biologist is so obvious that the failure to do so needs special explanation. (That radiation could be dangerous, that animals eat one another, were scarcely unknown facts.)

Attempting to explain this phenomenon, we are led into the second criticism of science, that it is "too abstract" and in virtue of that fact distracts attention from the actual complexities of the world around us which, so Marcuse once argued, it sets aside as "dirty" in comparison with the cleanliness of pure thought. (Plato is still alive and well.) One certainly has to grant, in response to this criticism, that science prefers to work with entities and relationships which are not the objects of everyday observation.

This comes out in the hierarchical constitution of science. At the top of the hierarchy stand mathematical physicists, who take as their ideal the discovery of such equations as $e = mc^2$, which make no reference to any particular entities. Next come those scientists who work with entities—genes, molecules, electrons—which are "theoretical" in the sense that we take them to exist and to behave in particular ways only as a consequence of accepting a set of scientific theories. Through a set of intermediate sciences which make use of such general formulae and such theoretical entities in the analysis of more accessible phenomena—earthquakes or disease—we pass finally to what is often dismissed as "natural history," the detailed description of phenomena in terms of their "everyday" properties.

(Note the way in which for a humanist like John Ruskin the hierarchy was entirely reversed: "Any science that adds to the descriptive knowledge of nature . . . is all to the good; any science that deals with analytic knowledge is only bad.")[5]

Thrusting aside the qualms of etymologists, let us invent the word "aristoscience" for the most prestige-earning kinds of science, without suggesting that the boundaries are sharp, that it is always an easy matter to distinguish "everyday" from "theoretical" properties. Aristoscience is specialized, analytic, abstract. And these characteristics are bound up with one another. It concentrates on a limited set of natural processes—specialization—and analyzes

5. From Robert L. Herbert, ed., *The Art Criticism of John Ruskin* (Garden City, N.Y.: Anchor Books, 1964), p. xv.

them, abstractly, into relationships between elementary constituents, relations which, preferably, are describable in purely mathematical terms.

Let us take for granted, against Ruskin and his like, the intellectual accomplishments of aristoscience. "Ariston," it must be remembered, is the Greek word for "best." They may not happen to interest or concern us, except where the technology derived from them impinges on our life. But if this is our attitude, we have a very limited appreciation of what human beings can discover and create. Our regret should rather be that aristoscience so often lies beyond our comprehension. One must still insist, however, that the scorn which aristoscience sometimes shows for "natural history" can be no less limiting. Geologists like Lyell, biologists like Darwin, physiologists like Harvey, psychologists like Freud and Piaget, sociologists like Marx and Max Weber have done at least as much, without the help of theoretical entities or mathematics, to transform our knowledge of ourselves and of the world as ever did Galileo and Newton and Einstein. (Freud is at his least illuminating when he tries to imitate the methods of aristoscience.) To put them at the bottom of an intellectual pecking order is merely arrogant.

If, furthermore, we are still ignorant about most of the phenomena we encounter in our daily life—whether it be human nutrition or the life history of animals—this ignorance can in part be set down to the aristoscientific emphasis on a very different kind of knowledge. Scientists themselves are beginning to emphasize as much. Even within physics—and remember Rutherford's dictum that "science is either physics or stamp-collecting"—the Cavendish Professor at Cambridge, Brian Pippard, has recently told aristoscientific physicists that it is time for them to turn their attention towards what he calls "the difficult and less elegant phenomena of the real physical world" in contrast with their past concentration on those phenomena the behavior of which can be described in beautifully concise formulae.[6] Somewhat less surprisingly, the biologist René Dubos has condemned the tendency of the aristoscientist "to become so much involved intellectually and emotionally in the elementary fragments of the system, and in the analyti-

6. A. B. Pippard, *Reconciling Physics and Reality* (Cambridge: University Press, 1972).

cal process itself, that he loses interest in the organism or the phenomenon which had been his first concern." This has the effect, he continues by pointing out, that "although everyone recognises that the very existence of natural phenomena and of living organisms is the manifestation of the interplay between their constituent parts under the influence of environmental factors, hardly anything is known of the mechanisms through which natural systems function in an integrated manner."[7]

One can see the effect of the hierarchical organization of science in the allocation of funds for cancer research. Environmental influences on the development of cancers can be *discovered* only by the observational methods of natural history, even if to *explain* them the scientist has to go beyond natural history. The aristoscientist prefers to investigate cancer through the study of cell structures. For he can then carry on aristoscience investigations which may win him scientific esteem however little they turn out to be related to the control of cancer. In contrast, an environmental investigation which fails to discover any correlation has little or no hope of being accepted for publication, none at all of adding to the scientist's repute.

This preference for approaching cancer through an investigation of the internal structure of the cell has sometimes been explained, by our conspiracy-minded age, in political terms; research workers, it is then argued, steer clear of environmental studies, since otherwise they could offend powerful industrial interests. It is much more likely, as I have suggested, that they prefer to work at molecular biology for reasons internal to the structure of science itself. Although the mere discovery by statistical investigations that, let us say, smokers are especially liable to contract a certain kind of cancer can suggest methods of control, it does not give us *understanding*. And that is what the scientist is looking for. He can justify this approach, in practical terms, in so far as it can lead to more far-reaching and efficacious methods of control; it would certainly be easier to remove from tobacco whatever makes it carcinogenic, if this is possible, than it is to persuade people to give

7. René Dubos, "Science and Man's Nature," in *Science and Culture*, ed. G. Holton (Boston: Houghton Mifflin, 1965), pp. 265, 264.

up smoking. But experience ever more strongly suggests that investigation into environmental factors, followed up by public health measures, is much more likely to be practically efficacious than further investigations into the cell—even if unlikely to win a Nobel Prize.

The cancer researcher's emphasis on cell studies is also a product of the fact that science has developed into one of the most highly specialized of pursuits, whereas the investigation of environmental effects leads us out of the laboratory and beyond the boundaries of any particular science into a variety of fields—chemical, meteorological, social as much as biological. This is precisely why physicists did not think it worthwhile to consult biologists about atomic explosions, or chemists to consult them about the biological effects of insecticides, or why biologists themselves for so long took such slight interest in ecology.

Specialization does not *necessarily* carry with it what Barzun calls "specialism"—a refusal to look around what one is doing, to examine its wider implications, its effects on systems or characteristics other than those which are being particularly examined. That flows rather from the attitude of an in-group, which turns its back on work which does not directly advance the profession, sometimes behaving, in Tolstoy's phrase, "like a deaf man replying to questions which nobody puts to him"—or, more accurately, nobody outside the in-group. But there can be no denying that specialization tends to create such an in-group, nowadays an in-group of such dimensions that the members of it can plausibly think of themselves as constituting a world in itself.

The emergence of science as a profession, along the lines Bacon envisaged, has proved to be, in this respect, at once a source of strength and a source of weakness. The scientist is a man with a career; no longer, as he once was, either an aristocrat, like Boyle, Cavendish, Lavoisier, or a maverick, like Priestley. He has to get results to keep his job. He is interested only in problems which, given sufficient diligence, he can reasonably hope to solve with the help of the general principles, the laboratory techniques, the co-workers he has at his disposal. "One of the reasons why normal science seems to progress so rapidly," as Kuhn puts it, "is that its practitioners concentrate on problems that only their own lack of ingenuity should keep them from solving."[8] The scientist is under no obligation to select problems merely because they are interesting or important to those outside the particular scientific community to which he belongs. If he does so, indeed, he is likely to find himself spurned by his coworkers as being no longer "one of them." It is he who will appear to be the deaf man, replying to questions which nobody has put. For they are not being put within the in-group.

There was a time when mathematicians were highly responsive to what went on outside mathematics, whether it was gambling, or insurance, or physics, or formal logic. But now, for the most part, they work at problems set by the inner development of the subject itself. The extraordinary tightness of scientific communities—the only *actual* example of what McLuhan called "global villages"—strongly discourages ventures into the outer world, let alone the suggestion that there might be something worth learning in that outer world. One sees how Paul Feyerabend can compare science to a church, closing its ranks against heretics, and substituting for the traditional "outside the Church there is no salvation" the new motto "outside my particular science there is no knowledge."[9]

To make matters worse, aristoscience breeds imitations, especially in the social sciences and the humanities. Philosophy, too, has come to take its problems from within itself, to concentrate on a "charmed circle" of problems, rather than to take them from science, or art, or religion, or society. Economists have been more interested in the mathematical refinement of highly abstract theories than in less elegant investigations into the operations of, let us say, mixed economies with their complex interactions between private and public consumers, private and public producers, monopolistic and free markets—the sort of society now characteristic of the industrial western world.

No doubt, it would be wrong to *blame* aristoscience for what, thinking of the way in which it apes aristoscience and despises "unscientific" reflection, we might call "snob-science"—at its

8. Thomas Kuhn, *The Structure of Scientific Revolutions*, 2d ed., International Encyclopedia of Unified Sciences, vol. 11, no. 2 (Chicago: University of Chicago Press, 1970), p. 37.

9. See particularly, Paul Feyerabend, *Against Method* (London: NLB, 1975). The motto, however, is of my own making.

worst in psychology and sociology, but infecting even scholarship. Aristoscience neither creates nor admires such sterile imitations of the real thing, in which it is supposed that to replace everyday words by signs for variables is at once to have made an enormous leap forward.[10] When the aristoscientist talks about social questions, one is often struck, rather, by his sociological naïveté, his refusal to believe that it takes work to find out what is happening in a society. But, as in the case of technology, he is in a way *responsible*, if not morally, then causally, even for what he despises.

Aristoscience is the principal source, too, of the emergence of that intellectual elite, living in a private world, immune from informed, as distinct from antiscientific, public criticism, to whose existence we have already drawn attention. Most of us quite literally do not know what theoretical physicists are talking about; if we think we do, it nearly always turns out that we have badly misunderstood. The aristoscientist, so much is clear, takes in his stride what seem to an outsider paradoxes so shocking—protons that move backward in time, let us say—as to demand reconsideration.

In short, the resemblance between the aristoscientist and the mediaeval theologian daily becomes more striking. Our everyday beliefs have as little relationship to what aristoscience tells us as did mediaeval theology to everyday mediaeval religious practice. I am not one of those who reach for their revolver when they hear the word *elite;* it is unreasonable to expect every scientific discovery to be comprehensible to everybody. But one need not be surprised, even if one deprecates the fact, that such remoteness should generate in relation to aristoscience, as in relation to some sorts of art and philosophy, a mixture of fear, terror, and hostility.

The word "aristoscience" has, indeed, rather more historical resonance than I realized when I first invented it. For I had forgotten what a rereading of Charles Gillispie's *The Edge of Objectivity* reminded me, that on his interpretation the French Revolutionary Terror struck down scientists "in a fit of vulgar, sentimental petulance against the hauteur of abstract science, the impersonal tyr-

10. Cf. Susanne Langer, *Mind: An Essay in Human Feelings*, 2 vols. (Baltimore: Johns Hopkins Press, 1967), vol. 1, chap. 2.

anny of mathematics, the superiority of the scientist over the artisan."[11]

Gillispie's interpretation of the Terror's relationship to science has been seriously questioned. But it so accurately reflects—as historians often do—an attitude of mind common in our own times that we read it with no surprise. "Hauteur," "impersonal tyranny," "superiority"—these our modern levelers certainly resent, and believe they find in aristoscience. Such critics will certainly not be assuaged by, for example, the physicist Bernal's picture of the world-to-come, a world in which scientists first become the government, and then emerge as a substantially new species. In the end, he suggests, the two groups would be obliged physically to separate; scientists will take themselves off to a satellite globe leaving behind the rest of the world's inhabitants as a sort of "human zoo, a zoo so intelligently managed that its inhabitants are not aware that they are there merely for the purposes of observation and experiment."[12] This—although Bernal makes no reference to the fact—is extraordinarily like the world Jonathan Swift envisaged in his *Voyage to Laputa.* And many antiscientists would see in it an only-too-plausible prophecy, the natural outcome of a situation in which control through understanding has come to mean, in practice, control by an elite.

The very success of science has been, from the point of view of its public stature, something of a handicap. Newton astounded Europe. Few could understand, as few can now understand, the mathematics which lie at the centre of his *Principia,* but his vision of the universe none the less excited their imagination, as did Galileo and Harvey before Newton. The impact of Lyell and Darwin in the nineteenth century was no less momentous. Robert Oppenheimer has argued that science cannot nowadays hope to have the same impact;[13] the discoveries which now startle scientists—discoveries about parity, let us say—are marvelous only

11. Charles C. Gillispie, *The Edge of Objectivity* (Princeton: Princeton University Press, 1960), p. 175.

12. J. D. Bernal, *The World, the Flesh, and the Devil* (Bloomington: Indiana University Press, 1969), p. 80. In the 1969 edition Bernal says that his final predictions need a little rethinking.

13. Robert Oppenheimer, "Physics and Man's Understanding," in *Knowledge Among Men,* ed. Paul H. Oehser (New York: Simon and Shuster in cooperation with the Smithsonian Institute, 1966).

in their eyes. Newton inspired the muse; Einstein has not done so. Indeed many a nonscientist would sympathize with Sir John Squire when to Pope's

> Nature and Nature's laws lay hid in night:
> God said, "Let Newton be!" and all was light

he added:

> It did not last: the Devil howling "Ho,
> Let Einstein be," restored the status quo.[14]

Shelley and Wordsworth both called upon the poet to mediate between science and the citizen. The Australian poet A. D. Hope complains that poets have failed to fulfil this function: "Poets have renounced their function and withdrawn from the intellectual adventure we call science."[15] But if, as Oppenheimer suggests, science has "got beyond" anybody but scientists, this renunciation is not a voluntary decision; the task of mediation is not one the poet can hope successfully to undertake. The contemporary German political theorist Jurgen Habermas maintains, indeed, that science now enters everyday human life not as an extension of man's *knowledge* of his world but only as a program for the technical control of it.[16] So the poet can write about the atom bomb—where science impinges, by way of technology, on daily life—but not about atomic theory which is divorced from that life. There are not, on this interpretation of the situation, "two cultures"; there is only one culture, a culture from which science is excluded except in virtue of its impact on that culture as technology. This is an extreme version of the view that science is now of no general interest except as technology, that it has quite lost its old function as a means of liberating man's spirit and enlarging his intellectual horizons.

If we look along the shelves in a book shop and see, where there once used to stand volumes on popular science, rows of occultism, mysticism, astrology, we might well conclude that Habermas is right, that science is no longer an ingredient of any significance in

14. Sir John Squire, *Collected Poems of Sir John Squire* (London: Macmillan, 1959).
15. A. D. Hope, Address at the conferring of degrees, Australian National University, 6 April 1972, as printed in the *Australian National University Reporter*, 14 April 1972.
16. Jurgen Habermas, *Toward a Rational Society*, trans. J. J. Shapiro (London: Heineman, 1971), chap. 4.

our general culture. But this would be an exaggeration. Science fiction has largely replaced popular science. It may leave everything to be desired as literature, it may be pessimistic, but it still serves as an intermediary between science and the general reader. Science journalists, if too few in number, serve a similar function in a more reliable fashion. Indeed, they are the source, often enough, through which scientists learn about developments in sciences other than their own. If fresh discoveries about parity created much scientific, but little popular, interest, the same cannot be said of recent work in genetics, immunology, radio astronomy. There never was a time, furthermore, in which the highest intellectual achievements were universally accessible. Even outside science, in philosophy, Plato's *Sophist* and *Parmenides* are as technical as any contemporary work, as abstract, as difficult to comprehend.

But as well as the *Sophist*, Plato wrote the *Republic* with its wide-ranging discussions of censorship, women's liberation, education, political control. The most audacious of all Greek thinkers turned his attention, like Aristotle after him, to problems which were of major concern to the society in which he lived. The rise of the aristoscience which Plato in that very dialogue first adumbrated has had the effect that our best minds no longer concern themselves with such questions, at least in a systematic way. They are left to third- and fourth-rate thinkers, or to superannuated scientists and scholars. To the major detriment of our intellectual life, the modern sophists, whether in sociology, in art theory, in education, or in politics are allowed to make their pronouncements without being subject to the sort of criticism Socrates and Plato directed against their own sophistic contemporaries. And something similar has been true within science itself; the best scientific minds have not devoted themselves to such humanly important topics as nutrition, ageing, or health-maintenance. There the cranks have been largely allowed to go their own way.

"Man does not live by bread alone"—that is one of the few things we know about human nutrition. To suggest that aristoscience ought to be abandoned just because it is by its very nature analytic and specialized would be to surrender one of our most remarkable achievements, one of the activities on which we can properly pride ourselves as human beings. Neither is it by any means the case that aristoscience is a mere intellectual game, formally beautiful but practically useless, that its analytical approach automatically makes aristoscience useless as a means of studying our complex, messy world. Aristoscientists, physics aside, have taught us a great deal about ourselves by studying our cells, our neurons, the chemical processes which go on in our bodies; with the aid of aristoscience we have reached a level of understanding, and a level of control, to which we could never have attained by clinical observations. One wants to insist, only, that the discovery of fundamental mechanisms, fundamental processes, is not enough; to understand why, to take an elementary example, an American sees a chicken but not a dog as food we need to do a great deal more than study his optical mechanisms, however such investigations can aid us in understanding how he can see a chicken at all. We have to embark, in a very broad sense of the word, on anthropological investigations.

More and more scientists, it is only fair to add, are coming to share this conclusion, that what, as the biologist Lord Ashby recently confessed, have been "regarded as second-class citizens in the hierarchy of sciences" must now, as he argues, assume a central place. But only too often, as we saw, these second-class citizens have been tempted to seek a higher social status by apeing aristoscientists. It is beneath their dignity to undertake the tedious descriptive work which is essential to any substantial attempt to come to grips with the effects of technology. A point has now been reached at which the intellectual endeavor of aristoscience, partly by attracting to itself so large a percentage not only of the best but even of the second-best minds, partly as a consequence of its scorning investigations which do not permit of elegant solutions, partly by giving rise, however unintentionally, to snob-science, tells against systematic inquiries of the highest quality into problems of deep human interest. (We are ignoring, at this point, its generation of technology.) So far, but only so far, the accusations of antiscientists have a degree of justification.

The third, closely connected charge leveled against science is that it encourages us, just in virtue of its analytic, abstract, approach, to think of the nonhuman world merely as a field in which human beings can exercise their power. And by so encouraging us, the objection continues, science helps to destroy the biosphere and, in the long run, the human beings who depend upon the biosphere for their survival.

How else might we regard nature, except as a field of action? We might, to take one possibility, think of it as quasi human, of every living thing in it—and perhaps not only of living things—as a sort of person, to be respected as such. That would certainly limit the degree to which we considered it proper to transform nature so as to make of it an instrument to serve our needs; there would be rights to be taken into account. One need not be surprised that such animistic doctrines are now, in certain quarters, experiencing a modest revival, as the only way of looking at nature, so it is said, which can justify ecological concern. But long before modern science got to work Christianity had rejected animism. And, I should say, rightly so; trees, let alone rocks, are not persons. They do not have aims, intentions; they neither love nor hate. We must look elsewhere than to animism if we are in search of a rational foundation for our resistance to ecological destruction.

Christianity left open another way of looking at nature: as a system of signs, a book which human beings could read, conjointly with the Bible, as teaching them what to do. This view is perhaps most familiar from Shakespeare's *As You Like It*—"tongues in trees, books in the running brooks, sermons in stones, and good in everything." But it was widely held. And science destroyed it. In destroying it, so a common argument runs, science at the same time weakened our sense of wonder and thereby made us more open to the suggestion that it does not really matter what happens to nature, that it can be transformed as we please, without loss, so as to suit the convenience of human beings.

A famous passage from Keats's *Lamia* makes the first of these points. (Remember that by "philosophy" Keats meant what we now call "science.")

> Do not all charms fly
> At the mere touch of cold philosophy?
> There was an awful rainbow once in heaven:
> We know her woof, her texture; she is given
> In the dull catalogue of common things.

The rainbow, he is saying, was once marvelous, awe-inspiring; now we have discovered what it is, it belongs "in the dull catalogue of common things." Goethe, in a similar spirit, objected to Newton's theory of colors. White was the symbol of simplicity, purity; to suppose it to be a combination of colors was to destroy its poetic essence.

There can be no doubt, as we have just said, that science is hostile to that conception of nature according to which it is sacred, improper to analyze, improper to control. Neither does it see nature, in the medieval and the occultist manner, as a set of symbols, a book for men to read—as palms are "read," or the disposition of leaves in a teacup. The rainbow is not, for science, a sign in the heavens. As long ago as the fifth century B.C., Anaxagoras became the first scientific martyr, exiled for declaring that the sun was no god, but a fiery stone. To the spirit of Anaxagoras, science must remain faithful. It cannot desist from exploring the moon merely because the moon was once a symbol of inviolate virginity. Yet is to analyze the rainbow in terms of refraction to reduce it to the "dull catalogue of common things"? It is not merely puerile to speak of the "marvels," the "wonders," of science. A twinkling star is rather a bore. It becomes much more of a marvel, a wonder, when we understand something of its nature, its history, even if it is no longer a "wonder" in the sense that Keats had in mind, no longer a miracle. No doubt the view that a rainbow is a message from God would make us reluctant to destroy it, if that lay in our power, and the same is true of the running brook. But the growth of science, as Skinner argued, need not limit our delight in rainbows, or our desire that our descendants should continue to delight in them. That is a good ground for not destroying them. And it leaves us free to transform what needs to be transformed without any superstitious fear.

But, the reply might come, is it not absurd to delight in a rainbow once we know that it is a mere appearance? And is this not precisely what science teaches us? There are deep puzzles here, which it would be wrong merely to thrust aside, even though we cannot explore them in depth. One or two vital points we have space to make. One is this: we can easily question whether electrons, or electromagnetic waves, or quantum leaps are anything but useful fictions—many philosophers of science have done so, some still do so. Science would be a good deal less interesting if they are right, but it would still survive. What *would* be fatal is the supposition either that the cracklings, the streaks of light, which you or I might observe as readily as the scientist in a physical laboratory, or the electron microscope, the radio-telescope, the cloud chamber, the scintillator which the scientist employs, are all of them fictional, mere products of the imagination. Yet these form part of our everyday world, in terms of which we describe them or, in the case of machines, tell others how to use them, ask that they be repaired or modified. Aristoscience cannot survive as an experimental investigation without relying upon our everyday tests for distinguishing between the real and the fictional, the actual and the hallucinatory, and without including in the class of the real and the actual what we, too, put there. In an important sense, indeed, science "leaves everything as it is." It still leaves us free to contemplate nature with enjoyment, with sensuous pleasure. For the scientist, as for the rest of us, dawn and sunset are still there to be admired; the scientist can still delight, according to the hemisphere, in the color of parrots or the song of the nightingale. If someone says: "But these are all, according to science, only appearances," this, quite certainly, is not in our ordinary, troubling, sense of "appearances." Discovering at Madame Tussaud's waxworks that what we took to be a guide was in fact a wax model, we cease to ask it questions. Such a model we might properly describe as a man only in appearance—as we might describe a transvestite as a woman only in appearance. It is not in any such sense, obviously, that the wife we love only appears to be a person and is, in reality, a cloud of electrons; we do not stop conversing with her as soon as we start reading theoretical physics.

Nevertheless, aristoscience does generate an attitude to the world which is peculiarly favorable to its technological transformation, peculiarly unfavorable to the contemplation of nature or its sensuous enjoyment. This arises from the fact that science, in its aristoscientific form, has a special concern: to look at the world *sub specie aeternitatis*. Neither the natural world of mountains and woods and rivers nor the species which inhabit that world have existed, or will exist, for eternity. From an astronomical point of view, they have had, and are destined to have, an extremely short history; it would be absurd, therefore, to introduce into physical laws any reference to human beings, to any other class of biological or geological objects, or to any phenomena which are dependent for their existence on the existence of human beings. These are all of them transitory, illustrations of laws, but not items in them. The lights and crackles in a laboratory are, from this point of view, while not illusions, not hallucinations, still not objects worthy of study in themselves; they are no more than a means which we employ to find out what the world is really like, a means which we are compelled to employ only because we are animals.

If we look at nature thus, then there is no intrinsic virtue in its taking one form rather than another. No doubt some natural objects are particularly pleasing to us as human beings. But for such aristoscientists as Poincaré or Russell the fleeting beauty of a rainbow is as nothing as compared with the beauty of a set of equations, representing, as those equations do, laws which will continue to hold long after rainbows and the human beings who delight in them have disappeared from the face of the earth.

Nature, as we encounter it in our daily experience, is, for aristoscience, ours to transform. We shall not, in so doing, destroy its beauty, for its beauty no more attaches to nature as it was than to nature after we transform it. Neither are we "destructive" when we cut down a forest. Our so-called destruction is more correctly to be thought of as a transformation. Properly understood, indeed, nature is a great storehouse of forces; there is no way of destroying such forces—the laws of conservation make that impossible. The typical natural object for Descartes, the typical human mind for John Locke—both of them writing under the influence of Galileo's science—is a piece of wax, flexible, malleable, ours to shape as we please whether by physical operations or by education, in the broadest sense of that word.

There is nothing in aristoscience, then, to call a halt to human destructiveness, there is much to encourage it. In rightly rejecting the view that the natural is a sign, a portent, a miraculous event, aristoscience has sometimes encouraged the view, as Augustinian Christianity also did, that it is unworthy of our serious attention, except as a starting point. But even although neither the forest nor the delight we take in it exist *sub specie aeternitatis*, it continues to be of the first importance to us, just because we are human beings. The destruction of a species is still the destruction of a species, for all that its energy is conserved. We have learned, under the tuition of science, not to be anthropocentric, not to suppose that the world exists for our sake, not to attribute to it characteristics which belong only to human beings or to what they create. But that does not mean that we should learn to despise, as unimportant, whatever matters only to human beings. Science itself is in that position. There is a sense, indeed, in which Wordsworth is right; to treat of things "as they seem to exist to the senses and passions" is to generate "truths as permanent as pure science." For the truths of science, as such, are as impermanent as the human beings who create or communicate them. But, of course, there is also a sense in which he is wrong; science sets out, as poetry does not, to discover relationships which hold eternally, as much when human beings are present as when they are absent. Admitting this, we still ought not to reject, as worthless, the world as our senses see it and passions feel it. For that is how we need to be able to see and feel the world if we are to survive, and enjoy our lives, as human beings. Entirely to reshape that world, to destroy what in it we find beautiful in the name of power over nature, is to give the preference to control over contemplation in a way which science may make possible but by no means necessitates.

To sum up the argument so far, we began by admitting that science can sometimes encourage a dangerous form of optimism, the search for unilinear "solutions" of social "problems," in a manner appropriate only to the solution of problems in physical science, the search for new ways of transforming the world on the assumption that what human beings can create is always superior to the processes it replaces. We have admitted, too, that science has been so structured as to emphasize analytic procedures even in cases where what is needed is the amassing of a great deal of concrete detail. But we are far from agreeing with Ruskin that aristoscience should be entirely abandoned in favor of natural history, that cells, neurons, molecules, genes, should be left in the darkness which once enveloped them. For again and again, it is through these abstract, analytical investigations that we have learned to understand and control the world around us. Often enough, we have however argued, such investigations have to be conjoined with studies of a different kind to give us full understanding, as the study of neurons has to be related to the biological and social history of the human species if it is to be of use in understanding human behavior. To understand why we see what we do see, and how we see it, we need not only optics but a knowledge of how human beings have had to evolve in order to go on living on the face of this earth. We have been prepared to grant to the antiscientist, too, that aristoscience can generate attitudes of mind which are socially very dangerous—a contempt for inquiries which are practically of the greatest human interest and which need the help of scientists even when their intellectual content is not of the highest order, a readiness to contemplate with equanimity forms of social organization in which the common man with his everyday problems is little more than a helot.

Chapter 4
Uniqueness, Imagination, and Objectivity

When Roszak condemns science as a "bewilderingly perverse attempt to demonstrate that nothing, absolutely nothing, is peculiarly special, unique or marvellous, but can be lowered to the status of a mechanized routine,"[1] he is summing up the accusations of a long line of humanistic antiscientists. To comprehend reality, they have argued, we must turn not to science—or not, at least, to aristoscience—but to literature and history. For reality, they point out, is made up of unique individuals, uniquely characterized, uniquely related, in unique situations. To generalize, they conclude, is to falsify. So according to Blake:

> Art and Science [i.e. knowledge] cannot exist but
> in minutely organised Particulars,
>
> And not in generalising Demonstrations of the
> Rational Power.[2]

Blake was an enthusiast. But even so sober a thinker as the English psychologist-philosopher James Ward was prepared to argue that only history offers us facts, that physics confounds abstractions with realities.[3] Not only knowledge but morality, so it is also maintained, suffers in consequence. "Evil," according to Sartre, "is

1. Theodore Roszak, *The Making of a Counterculture* (London: Faber and Faber, 1970), p. 229.
2. William Blake, *Jerusalem* f. 55, lines 62–63.
3. James Ward, "Mechanism and Morals," *Hibbert Journal* 4 (1905): 79.

the systematic substitution of the abstract for the concrete."[4] And that substitution is precisely what Sartre's younger contemporary, the philosopher of science Gaston Bachelard, had taken to be science's task.

It is not difficult to understand these sentiments. For it *is* true both on the one side that we ourselves and everything around us are, in a certain sense, unique individuals and, on the other side, that with rare exceptions—like cosmological speculations about "the beginning of the universe" which, even so, many would wish to expunge from science as having no proper place there—science is not interested in individual events, for and in themselves, but only in what repeatedly happens. This is true not only in aristoscience but even at the level of descriptive biology and geology. The "life-histories" constructed by biologists are not biographies. A biographer emphasizes the peculiarities of a particular human life, as compared with other human lives; the more abnormal the life is, the more extraordinary, the more likely it is to attract the biographer's attention. The natural historian's concern, in contrast, is with the species; an abnormal member of the species he casts aside as a mere "sport." Similarly, for a sociologist who takes social mobility as his theme, Bill Smith is of interest only in so far as the style of his social progression to becoming William Perkins-Smith is in some way characteristic. If, "by a set of curious chances," he is precipitated into being Lord Koko of Titipu, he at once attracts the attention of the biographer but will be dismissed as "uncharacteristic" by the social scientist. (The historian occupies an intermediate position in this respect.) A painter may pick up a particular stone on the shore, or a particular leaf in autumn, and be fascinated by its unique color and form. But that is not the scientist's way.

At the level of aristoscience, furthermore, all reference to everyday species, let alone to individuals, vanishes; its equations describe, not the peculiarities of individuals or individual species, but functional variations between properties. When this is not true, when, for example, nuclear physicists are compelled to engage in natural history—not to be sure the history of everyday species but

4. Jean-Paul Sartre, "Saint Genet," in *Against Interpretation,* Susan Sontag (New York: Farrar, Strauss and Giroux, 1966), p. 97.

of distinctly "un-everyday" particles—many physicists, Heisenberg for one, are discomfited and call for a halt in the name of a purely geometrical physics.

Yet none of these considerations is sufficient to demonstrate that science falsifies. If the uniqueness of things, as the more metaphysically intransigent humanists would have us suppose, were of such a kind that any description applicable to any particular thing was automatically inapplicable to anything else, then certainly science could in no way refer to the world—it would be an elaborate game with man-made fictions. But then, too, biography would collapse along with physical science. For the biographer can construct his story about, let us say, Julius Caesar, only by making use of descriptive terms which have a wider application, terms which are equally descriptive of other men and women who loved and were hated, wrote history, or assumed authority. *Hamlet*, to take the kind of example to which humanists point as the prototype of uniqueness, is certainly unique, in the sense that there are many things which can be said about it but not about any other play. Nonetheless it can be quite accurately described as a Jacobean revenge play. And what is true of *Hamlet* is true, *mutatis mutandis*, of everything else. Some things, like *Hamlet*, are extraordinary, some, like fingerprints, commonplace, but each and every thing is unique, in the sense of being distinguishable from anything else, and not unique, in so far as it is describable in the same terms as some other things.

Just because nothing is unique in the sense that all its characteristics are peculiar to it, science can help us to understand individuals even although it does not mention them. Of course, we have to rely on our judgment in determining whether any particular individual event "comes under" a particular scientific law or is of a particular type. A psychotic classification system does not of itself tell us that Jones is a manic-depressive. But it does tell what we have to find out about Jones in order to determine whether he is a manic-depressive. Although it does not enable us to anticipate in fine detail the character of Jones's fantasies or to determine the etiology of his disorder, it can still help us to understand Jones and, for all that he is a unique individual, to cure him. *Jones is a manic-depressive* "falsifies" only if we wrongly suppose it to be an attempt to tell us *everything* about Jones. But to read it as "the only characteristics Jones possesses are those which he shares with all manic-depressives" is simply to misread it. Science does not *replace* the concrete; it helps us to understand it.

The humanist and the scientist can easily lose patience with one another. The humanist devotes his life to studying, let us say, the decorative devices on South Italian vases of the fifth century B.C. He could not care less, as likely as not, about Chinese vases or modern ceramics, let alone about molecules. (Scholars are, in general, more specialized than scientists and the worse for it.) Science may provide the scholar with techniques; the scholar may even boast that he approaches his work "scientifically." But what comes out of his investigations is not in the least like science, even at its "natural history" level. Such a scholar may find it as hard to understand how somebody could devote his life to the search for functional relationships between extremely general properties as the aristoscientist finds it hard to understand how anyone could be devoted, except perhaps as a hobby, to dating particular vases or allocating them to a particular painter or a particular workshop. As for the humanist's interest, which he may use the vases to illustrate, in how it felt to be alive in a particular place at a particular time, this to the aristoscientist is positively indecent in its anthropocentrism.

But each of these forms of inquiry can offer its own kind of understanding. It is important to recognize how differently people can feel, how differently they can see things, if only to destroy those rigid expectations which take shape as bigotry or as an incapacity to understand why "everyone can't be like me!" That a biographer is not a sociologist is an objection neither to biography nor to sociology; that a family doctor is not a pathologist is an objection neither to family doctoring nor to pathology. If humanists or family doctors can properly feel resentful it is only in so far as they find themselves looked down upon by the aristoscientist as second-rate citizens in the republic of learning; if scientists can properly feel resentful, it is only when they are spurned as Philistines by those whose minds turn always to distinctions, never to identities. The mere consideration that things are different from one another gives us not the slightest ground for believing that general equations have therefore no relevance to them; the mere consideration that such equations are general in their application entails neither that the differences between the things they apply to are only "appearances" nor that they are unworthy of serious investigation.

But we have still to consider Sartre's claim that to think abstractly is at once to fall into evil. One sees what he means. As every war makes only too plain, it is much easier, without compunction, to deceive, mutilate or kill a man or woman whom one can think of, simply, as a German, a Roman Catholic, a Jew, a Muslim, a Communist, a counter-revolutionary, than as an individual human being. (Under these circumstances, we *do* suppose that "he is an *x*" says all that can be said about him.) And science, the argument would run, is at fault not because it is *intellectually* wrong to think of entities in general terms, but because to think of them thus is *morally* wrong, not only when they are people but even when they are things. When the aristoscientist tells us with satisfaction that every member of each particular class of elementary particles is strictly identical—so that, if he is right, elementary particles are an exception to what I said above about uniqueness—and is dissatisfied only because there are many such classes, he is betraying an attitude of mind which, so such critics maintain, cannot but spread to his attitudes to persons, to the depreciation of individual differences as something which ought as far as possible to be planned out of existence. "Scientific planning," "scientific socialism," "scientific management" all display this hostility to individual differences.

Replying to such critics, one cannot truthfully maintain that such an indifference is never to be found among scientists, that there are none of them to whom democracy, with its emphasis on diversity, on individual rights, is unendurably disorganized, or even that their hostile attitude to democracy is totally disconnected from their aristoscience. J. D. Bernal is an example to the contrary, and only one among many. Teilhard de Chardin was by no means merely eccentric when he wrote "I needed to know that 'Some one Essential Thing' existed to which everything else was only an accessory or even an ornament" or that "the multitude of things is a terrible affliction." Neither is he wholly exceptional in the heartlessness, the hostility to diversity, which runs through his work.[5]

Teilhard, however, was not an aristoscientist. Scientists with

5. For a defence of this interpretation, see John Passmore, *The Perfectibility of Man* (New York: Charles Scribner's Sons, 1970) p. 255.

paleontological interests are often fascinated by diversity; their generalizations are of a low level. And the attitude of mind Teilhard expresses, if it is relatively common among aristoscientists, is equally common among philosophers, mystics, socialists, and not unknown—as "minimalist art" even among artists. The monistic-totalitarian impulse preceded science and will survive it.

It is certainly worth keeping in mind that science does attract men and women who are dominated by such impulses, worth keeping in mind, especially, if they are likely to occupy positions of power. But if to be a scientist is not necessarily to be devoted to personal freedom or ecological diversity, it is not necessarily, either, to be devoted to "scientific socialism," "scientific management," and their fellows. Like snob-science, these phrases simply "cash in" on superficial resemblances to science in order to win wider support. (Now that the image of science is tarnished, "humanistic socialism" is replacing "scientific socialism" as a propagandist's catchword.) There are certainly scientists, and many of them are aristoscientists, who are as little interested as the most commercially minded "developer" in the preservation of ecological diversity, as little interested as the most ardent totalitarian in cultural diversity. Science-based inventions, one must also confess, can be used by the state to *enforce* uniformity. But in neither of these respects is science exceptional, even in contrast with the humanities. Poets, as we have already seen, are not always devoted to liberty; totalitarian societies use art, philosophy, religion as much as technology to strengthen their power.

The institutions of science, furthermore, if often hierarchical are by no means dedicated to the destruction of individual differences. Indeed, they are not uncommonly condemned as "elitist." Scientific academies, scientific selection committees, grant-giving bodies, all pride themselves, and not unjustly, on being able to distinguish between work which is, in Roszak's phrase, "special, unique or marvellous" and mere hack-work, science "lowered to the status of a mechanised routine." If one fears for the long-term fate of science, now that it has been so largely taken over by industry and by the state, this is precisely because both state and industry tend to find science disorderly, in the liberty, the degree of individuality, it allows to the original researcher. In authoritarian states, the scientist may often be found at the centre of dissent, quite as much as the artist and much more than the scholar. (Although appearances may be deceptive; the scientist is far more likely to be left at liberty than the scholar.) And many of us would see more ground for hope in the liberal dissent of a Sakharov than in the fanatical dissent of a Solzhenitsyn, however highly we might estimate Solzhenitsyn's gifts as a novelist.

It would be paradoxical in the extreme, then, to call a halt to science in the name of the "unique individual." With Norbert Wiener, we may lament "the degradation of the position of the scientist as an independent worker and thinker to that of a morally irresponsible stooge in a science-factory."[6] But what we are then lamenting is the decline of science in a consumer's society, not something inherent in science as such. The most the humanist can properly do is to enter a caveat to the presumption that, intellectually speaking, nothing but aristoscience is of any value, that a fascination with diversity is mere "stamp-collecting," involving no serious intellectual effort. Such a mental attitude aristoscience has certainly encouraged, and it has been particularly devastating in so far as snob-science has sought to imitate it, however crudely, whether in literary criticism or in history or in education. To us, as human beings, diversity *is* important; as are such unrepeated events as the Heian period in Japan, the French Revolution, such unrepeated objects as Shakespeare's *Hamlet* or Michelangelo's *Moses* or Newton's *Principia*, on which humanists concentrate their attention. But to say as much is by no means to deny that the discovery of uniformities can be both theoretically fascinating and of great practical significance.

We may reasonably mistrust those who are wholly dedicated to the discovery of very general principles, however, if they try to talk to us about our social and political problems. Even in discussing ecological problems, where complexity, diversity, is so essential to understanding, scientific publicists tend to seize upon what they take to be "the one thing needful," whether it be, in Ehrlich's manner, population control or, in Commoner's manner, control over chemical pollution. And in relation to every social problem, there is not one thing needed but many things needful. The math-

6. Norbert Wiener, "A Rebellious Scientist After Two Years," *Bulletin of the Atomic Scientists* 4 (November 1948): 338-39. Quoted by Lewis S. Feuer in *The Scientific Intellectual* (London and New York: Basic Books, 1963), p. 399.

ematicians and musicians, the sages of *Gulliver's Travels*, had by them a servant armed with a balloon with which he struck them at intervals to recall them to the complexities of everyday life. Perhaps we need such servants in our highest science councils. The scientists, all the same, should not be struck too often—and with a balloon, be it remembered, not with a bludgeon.

An even more common charge against science is that it "destroys the imagination." At first sight, nothing could be more absurd. Compared with theoretical physics, *Alice in Wonderland* is a naturalistic novel. As C. S. Peirce argued, "next after the passion to learn there is no quality so indispensable to the successful prosecution of science as imagination."[7] Kuhn, no doubt, has drawn a sharp distinction between what he calls "normal science" and the sort of imagination-rocking science, relatively rare, which takes shape as a scientific revolution. In general terms, he is undoubtedly right. Science which fully exerts the imagination is as rare as art which does the same thing. Most art, one might say, is in Kuhnian terms "normal art," creating minor variations on the accepted forms of the day, working out the possibilities inherent in those forms to the point of exhaustion. This is as true of poetry as it is of music and painting. If one has to confess that much science is boring and many scientists are boring people, the same is true of any form of human activity; much poetry is boring and many poets are boring people.

There is, no doubt, a special peculiarity of science as compared with, let us say, art or poetry. In the nineteenth century the Germans finally put into practice the Baconian ideal of team research, of discovery through perseverance. For the first time, men with second-rate minds—worthy, reliable, but uninspiring—could make for themselves a career in science. They constitute Norbert Wiener's "stooges." Modern science is unthinkable without their aid. They are no more imaginative, if no less imaginative, than the bulk of historians, philosophers, scholars. This is the sort of scientist that Roszak is particularly attacking. But their presence within science should not obscure the fact that science is still one of the great triumphs of the human imagination.

7. C. S. Peirce, "The Scientific Attitude," *The Philosophy of Peirce: Selected Writings*, ed. Justus Buchler (London: Routledge, 1940), p. 43.

If its critics fail to recognize its imaginativeness this is, often enough, because they misunderstand the nature of imagination.[8] Sometimes they identify it with the capacity for forming vivid imagery. It is then concluded that drugs—or the troglodytic practices of Tibetan monks—enlarge the imagination because they increase the range, or intensify the vividness, of imagery. But there is no necessary connection between imaginativeness and the capacity to image vividly. The question whether Galileo, or Newton, or Darwin was imaginative is certainly quite independent of the question whether their imagery was vivid. Many of us, I suppose, under conditions of intense fatigue, or illness, have experienced an intensification of imagery. It is not the least like those imaginative flashes of insight we have less often experienced.

Neither is imaginativeness the same thing as fantasy. I am not being imaginative—on the contrary, I am being completely unimaginative, completely conventional—when I imagine myself, in a fantasy, as conducting a symphony orchestra or winning a Nobel Prize. Fantasy, in this sense, is simply the capacity for lying to myself about who I am and what I can do; as such it has no virtue. One's fantasies, one's lies, *can* be imaginative but they are for the most part unimaginative. Science criticizes fantasies, and that is what its critics cannot bear. But it does so in an imaginative way.

The nature of imaginativeness comes out more clearly when we contrast imaginative thinking with routine thinking. Routine thinking is conventional in character, in its goals, in the methods it employs, the principles it invokes, as when, for example, I find out when my plane leaves by consulting a timetable or calculate the area of my land by multiplying its breadth by its length. To be imaginative, in contrast, is to discover unexpected and illuminating relationships, to invent new explanatory principles, to discover familiar principles in an unfamiliar context. Shakespeare's metaphors are imaginative and so was Darwin imaginative when he saw the influence of natural selection in the most unexpected places, or Galileo when he amalgamated celestial and terrestrial mechanics.

8. I have discussed imagination in art and science a little more fully in John Passmore, *Art, Science and Imagination* (Sydney: Sydney University Press for the Australian Academy of the Humanities, 1975).

Looked at thus, the imaginativeness of science is at once obvious. Then why should it be so often condemned, not only by Roszak, as unimaginative, mechanical, routine? In part, the scientists themselves are to blame. Intent on emphasizing that experimental method which is their badge of authority, their professional distinguishing mark, they have deprecated their own imaginativeness. They have wrongly supposed—exactly like their critics—that the exercise of the imagination and the experimental method are essentially opposed to one another: they have failed to understand that science is the wedding of the two. But, as well, science has suffered from its successes. By exercising their imagination, scientific innovators have made it unnecessary for the beginning scientist to exercise his imagination. Very usefully, they have discovered an immense variety of principles and practical procedures, which can safely be adopted as a mode of solving problems. To take a very elementary example, if the young scientist wants to know whether a solution is acid or alkaline, he does not have to think out how to determine the difference, he simply reaches for the litmus paper; if he wants to know the temperature, he consults a thermometer. So in a way Roszak is right; science does make procedures routine which at one time involved an exercise of the imagination. And if science is badly taught, students can be left with the impression that all it does is to use these routines—if his teacher does not make it apparent to them, that is, that these routines were at one time the product of imaginative responses to problems.

As science is encountered in the schools, it is often as a matter of learning principles and applying them by routine procedures. It is learned from textbooks, not by reading the works of great scientists or even the day-to-day contributions to the scientific literature. At the university level, indeed, this is still largely true. The beginning scientist, unlike the beginning humanist, does not have an immediate contact with genius. Indeed there is some evidence to suggest that school courses can attract quite the wrong sort of person into science—unimaginative boys and girls who *like* routine—while at the same time deterring the imaginative; evidence, that is, that the conventional school science course was a useful apprenticeship only for the second-rate scientist—as well as wholly concealing from the nonscientist the imaginative quality of science.[9] But such

9. See Liam Hudson, *Contrary Imaginations* (London: Methuen, 1966), p. 60.

defects in traditional scientific education, if they help to *explain* why humanists, most of them acquainted only with science in its most elementary form, can condemn it as unimaginative, still do nothing to *justify* that judgment.

Stung by humanistic critics, scientists sometimes counterattack by maintaining that there is no room for the operation of the imagination except in science. Consider, for example, the atomic physicist I. I. Rabi's now-notorious dicta that the plays of Shakespeare are only "wonderful, glorified, gossip" and that "the imagination that deals with the mystery of life" is only to be found in science.[10] These are not, I think, the isolated judgments of an eccentric physicist; they give open expression to sentiments which are quite widespread, if seldom so openly expressed. But they are quite as foolish as the humanistic countersentiments. Science did not *invent* the creative imagination; indeed, it was itself the offspring of that imagination.

Of course, one must hastily add, neither science nor scholarship is *simply* an exercise of the imagination. If Peirce saw that imagination was essential to science, he at the same time warned us that the scientist "cannot prosecute his pursuit long without finding that imagination unbridled is sure to carry him off the track." It is in its manner of bridling, not in its being imaginative, that the peculiarity of science consists. Antiscientists sometimes admit that scientists can have a capacity for "dazzling inspiration." But they still condemn science on the ground that for the scientist "taste, inspiration and intuition do not *prove* anything."[11] For science *tests* its intuitions and inspirations; it hopes to arrive at public knowledge, the claims of which anybody with sufficient skill can examine for himself.

Science, in consequence, is hard work, as having inspirations is not. There is an old mystical saying that "man was born to find the truth without labour." That no scientist believes; he may, to be sure, have a sudden inspiration, but that is only the beginning, not the end, of his task. Here we have come up against another point at which, according to the more extreme antiscientists,

10. I. I. Rabi, "The Interaction of Science and Technology," in *The Impact of Science on Technology*, ed. A. W. Warner, Dean Morse, and A. S. Eichner (New York: Columbia University Press, 1965).

11. Theodore Roszak, *Where the Wasteland Ends* (Garden City, N.Y.: Doubleday, 1972), p. 154.

science belittles man; the "godlike" intellect which Darwin ascribed to Galileo and his successors is not, in their judgment, truly godlike *precisely because it does not offer us a direct illumination of the world by immediate insight.* But the great merit of science, one may reply, lies in its recognition that when the question at issue is what the world is like, immediate insight is just not enough, that man's intellect is *not* godlike.

If, indeed, science cannot bring its theories to the test of observation and experiment, as distinct from relying on direct illumination, inspiration, it simply does not exist as a unique form of human culture. Its distinction from myth, from art, from occult fantasies—none of which, unlike science, is a peculiarly modern or Western creation—depends upon its capacity to undertake this task, just as the peculiarity of mathematics lies in its capacity to prove. Anxious to ensure its distinctiveness, discerning in its critical spirit both its moral and its intellectual virtue, philosophers of science, including scientists in their philosophical moods, have sometimes exaggerated, however, the extent to which such tests can be final. And this exaggeration, too, has now created problems for the public image of science, both at the popular and the theoretical level.

Encountering science, for the most part, at such fringes as medicine or nutrition or ecology or cosmology, the general public is puzzled to find "expert opinion" so markedly divergent. What they learned about science at school persuaded them, if it persuaded them of anything, that science works by applying formulae which have either been deduced, in a Euclidean manner, from axioms or generalized from attested facts by making use of inductive methods. At the very least, they were persuaded that scientists have ways of showing, rapidly and confidently, that a hypothesis is false. In so far as they were attracted by science, what attracted them, often enough, was the security which it offered. Disillusioned by confident pronouncements and no less confident counterpronouncements about, let us say, diet and heart disease, supersonic flight and the ozone layer, industrial development and the heating up, or cooling down, of the atmosphere, big bangs and perpetual creation, they now turn against science as a mere pretender.

Such a response a better understanding of science can largely dispel, as well as an understanding of its limitations, of the extreme difficulties it encounters in analyzing complex situations involving feedback. Much more fundamental is a scepticism directed not against this or that scientific pronouncement but against the viability of the whole scientific enterprise insofar as that depends, or so I confidently asserted, on the possibility of bringing hypotheses to the test of observation and experiment. To put what is substantially the same point differently, the antiscience sceptic seeks to overthrow the concept of *objectivity*, at least as I should understand that concept. "The very reason for the success of science in replacing religion—its seeming objectivity and universality—is a chimera," so one such sceptic has recently written.[12] And in his *Against Method*, basing his argument on a demythologized history of science, Feyerabend has defended an anarchistic conception of science, according to which "anything goes" and science is, indeed, but one variety of ideology.[13] We must now, then, seek to rebut the last of the charges I began by listing: that objectivity is either mythical, or a mask for callousness and political opportunism.

Let me make certain concessions at the outset. First of all, as will be sufficiently apparent from what I have already said, to describe science as "objective" is not to claim that scientists are always right or even, what is a somewhat weaker claim, that they go wrong only when they diverge from the established method of getting things right. Both claims are wholly untenable. There is no established method, whether it be inductive or deductive, of getting things right. There are only methods which will make it less likely that the scientist will commit certain kinds of error.

Secondly, the scientist never ceases to be a human being, with human passions, human weaknesses. He can in no way avoid what Polanyi calls the "passionate, personal, human appraisals of theories."[14] Insofar as the scientist supposes his objectivity to consist in his having an intellect godlike in its freedom from human passions, he is quite mistaken. Commitment is as essential to success in science as it is to success in philosophy or in art.

Thirdly, the scientist cannot somehow shake himself free from

12. S. H. Shuman, "Scientific Manipulation of Behaviour," in *Equality and Freedom*, ed. Gray Dorsey, vol. 2 (New York: Oceana Publications, 1977), p. 699.
13. Paul Feyerabend, *Against Method* (London: NLB, 1975), pp. 307–9.
14. Michael Polanyi, *Personal Knowledge* (London: Routledge and Kegan Paul, 1958), p. 15.

his intellectual milieu, from what Collingwood called the "presuppositions" of his times or Kuhn "scientific paradigms." A scientist is a social being, working in a particular scientific community, within a broader economic and political society. Every now and then a great scientist, or a great philosopher, draws our attention to how much we have been taking for granted. But he, too, is a member of such a community, not an absolute "outsider." However much he sees through, he always sees with the eyes of a human being, born at a particular time in a particular place.

Fourthly, to speak of the scientist as "bringing theories to the test of experiment and observation" does not imply that he is ever in a position to set his theories against "pure data," if by "pure data" we mean something like the "sense-data" of classical British epistemology, direct records of experience, freed from any sort of any risk of error or any theoretical preconceptions. Some contemporary philosophers of science, reacting against sense-data or "protocol sentences," have, in my judgment, gone too far in asserting that even our everyday common sense judgments incorporate a "theory"—the word "theory" is in such contexts unendurably stretched. But it is certainly true that what the *scientist* thinks of as his "data" are not only corrigible in principle but for the most part theory-dependent. Even to accept the readings of a thermometer as recording changes in temperature is to assume the theory that mercury expands with heat in a steady fashion; Galileo's opponents were not merely obscurantist in refusing to count what they saw in the lens of a telescope as an observation. As science develops, this theory-dependence comes to be more and more obvious.

In essence, I would still stand by what I wrote a quarter of a century ago on this theme, with special reference to David Hume but with a broader intent:

> If, in the attempt to keep science "pure" and "rational," we try to exclude the effect of special interests, prevailing modes of thought, the desire to get on with a job, we destroy its vitality; and we are confronted with a mechanism of confirmation which seems quite inadequate for the task it has yet so successfully performed. Hume's successors thought he had shown the need for formulating "an inductive logic"; but what he really showed is that there is not such a logic. He was wrong in concluding that logic played no part in scientific reasoning—wrong in thinking that science must be either

demonstrative or inductive—but he was right in insisting that scientific thinking cannot be described as a simple application of logic. His theory of the imagination is crude and misleading, but his insistence upon the importance of the imagination as a copartner to observation at every stage in our thinking contains a lesson we have yet fully to appreciate. A crude positivism, a phenomenalistic "empiricism," is quite alien to the spirit of Hume's philosophy.[15]

If, indeed, we compare actual science with an ideal Cartesian-type science, deducing its conclusions from indubitable axioms, or an ideal Mill-type science grinding out general principles by the application of "inductive methods" to pure data, or even an ideal Popper-style science, made up of bold conjectures and decisive refutations, we can easily persuade ourselves, as Feyerabend has finally done, that it has no better claims to be accounted objective than has religion.[16] But we have only to compare science with other ways of speculating about the world to see how actively it seeks public knowledge—"universal and objective"—as little influenced as can be by personal advantage, political ax-grinding or simple wish fulfilment.

Let me try to justify this judgment by contrasting scientists with prophets. First in style, a matter of no little importance, as the founders of the Royal Society fully recognized. Scientists present their "observations"—in the broadest possible sense of that word—for criticism by their peers, in a language which is as precise as possible, designed to facilitate so far as can be the critical consideration of what they have to say. The prophet, in contrast, is cloudy and evasive, preferring parable and analogy to direct statement, relying for much of his power to win converts on the fact that his "message" lends itself to a wide variety of interpretations. Secondly, in their reaction to criticism. The scientist is only human; criticism may deeply disturb him. Nonetheless he expects and invites it. The prophet, in contrast, damns his critics as "men of little faith." Thirdly, the criticism to which the scientist subjects himself is, to a striking degree, internationalized so that it is in

15. John Passmore, *Hume's Intentions,* 2d ed. (London: Duckworth, 1968), p. 158.

16. Paul Feyerabend, "How to Defend Society against Science," *Radical Philosophy Eleven* (1975). On this theme see Israel Scheffler, *Science and Subjectivity* (Indianapolis: Bobbs Merrill, 1967) and the essays in *Criticism and the Growth of Knowledge,* ed. Imre Lakatos and Alan Musgrave (Cambridge: University Press, 1970).

some measure freed from the limitations imposed by national prejudices and national traditions. It extends, too, over time so that errors arising out of the "mental sets" of a particular period can expect to be in the long run corrected. A prophet, in contrast, may be internationally *interpreted,* but tends to be merely ignored by those who are not his followers—or inveighed against, as distinct from being subject to close criticism. Fourthly, although both scientist and prophet make predictions about the future, predictions on which their right to be heard with respect are partly based, the scientist's predictions are founded on independently criticizable principles, the prophet's on his own mere asseveration or his interpretation of texts so deviously expressed that they are subject to no precise canons of interpretation. Fifthly, as Popper long ago pointed out, the predictions of the scientist are so temporally and spatially specified that we can have ways of determining whether or not they are successful; the prophecies of the prophet are expressed in a manner designed to ensure that whatever happens they are still not falsified. (Characteristically, they contain such expressions as "in the fullness of time" or "in the long run" or apocalyptic symbols.)

Looking thus at the procedures of scientists, comparing them not with the conceivable procedures of a rational automaton but with those adopted by prophets, sages, poets, and transcendental metaphysicians—scholars are more like scientists, although they do not attempt to predict[17]—we can readily see how much more properly they can be labeled "objective." The institutional life of science is designed to submit the "observations" put before it to rigorous international and cross-temporal criticisms; it demands that these observations take a form which will allow of such criticism. The scientist tries to repeat "observations" under different circumstances, in different laboratories; he deliberately seeks out exceptions, or previously unenvisaged instances where the "observations" can be brought to the test; he seeks to show that new "observations" are deducible from already-accepted principles—or inconsistent with them. None of these methods is

17. Compare on this theme John Passmore, "The Objectivity of History," *Philosophy* 33 (1958): 97–111, reprinted in W. H. Dray, ed., *Philosophical Analysis and History* (New York: Harper, 1966) and in Patrick Gardiner, ed., *The Philosophy of History* (Oxford: University Press, 1974). For the structure of the scientific community see Jerome R. Ravetz, *Scientific Knowledge and Its Social Problems* (London: Oxford University Press, 1971).

infallible; the scientist cannot simultaneously bring to the test everything he believes; he can be mistaken in believing that this rather than that is what he ought to be rejecting. But we are not evincing a blind faith when we judge that the scientific methods of criticism are the best methods we know of, at least when what is at issue is our detailed knowledge of the world, certainly better than a reliance on what Bacon called "the first offers and conceits of the mind." These critical methods, originally based, of course, on the self-critical practices of a very few geniuses, are "internalized" by the vast majority of scientists. But science does not take it for granted that this will be so; it does not relax its vigilance.

I do not mean, of course, that we should refuse to listen to those who are not scientists, on the ground that what they tell us has not been subject to so rigid a scrutiny. For one thing, in areas of the greatest human consequence science may have nothing to say; for another thing, as we have already seen and as recent critics of science have been at pains to emphasize, the processes of criticism may be infected by assumptions of a sort which are so prevalent that they are not detectable from within the science they infect. New developments inside a science have not uncommonly been influenced by ideas which have arisen outside science, as by Einstein's reading of philosophy. Neither need the scientist's reading, even, be of such an intellectually respectable kind, subjected to critical processes of a different, but no less rigorous, sort.

Recent investigations into the history of science have helped us to see more clearly what an important role quite wild speculations can play in scientific "breakthroughs"—the speculations, let us say, of German "natural philosophy" or cosmic evolutionism or Pythagorean number mysticism. We can, indeed, welcome Feyerabend's "anything goes" if we interpret it as asserting that society at its most creative will contain a great variety of amorphous ideas, that if such ideas were to be expelled as unworthy of our modern times, science might become moribund, collapse into trivialities or, at best, become a mere handmaid of technology. But the fact remains that such speculations, important and interesting as they can be, do not supply us with "objective knowledge" in the sense that science does. And I do not think that anything need be done positively to encourage them, that they ought to be left uncriticized or that they are ever likely, as science advances, to die out. Granted a modicum of social and political freedom, one need not fear for their future. Science, just in virtue of its quest for objectivity, lives much more precariously, in a world in which the spirit of objectivity represents a constant danger to ideology.

It will be obvious, however, that to speak of what passes scientific tests as "objective knowledge" is to assume that the scientific community, if sometimes narrow and blind, is in general worthy of our confidence, that we can trust scientists to be rigorous and unsparing critics. But are we entitled any longer to have confidence in the scientific community? Feyerabend has been moved to describe the contemporary scientific community thus: "Good payment, good standing with the boss and the colleagues in their unit are the chief aims of these human ants who excel in the solution of tiny problems but who cannot make sense of anything transcending their domain of competence."[18] Can we honestly reject this description of the scientific, or indeed the scholarly, scene as nothing but a malicious libel? And if not, can we expect from such servile insects a scrutiny which will be fair and just and thorough?

This question: "Why should we place any particular confidence in the judgment of the scientific community?" was one that, until very recently, was scarcely raised. It was generally assumed that scientists were human beings of exceptional honesty and exceptional openness, engaged in a disinterested manner—not, of course, an *un*interested manner—in the pursuit of truth. One could not possibly account in any other way, indeed, for their devotion to a form of life which brought with it so little in the way either of actual worldly rewards or anticipated heavenly recompense. And, in consequence, one could safely extol the critical scrutiny of scientific "observations" by the community of scientists as being as near to an ideal scrutiny as one could hope to achieve, limited, no doubt, by human ignorance, by human conservatism, but certainly not distorted by considerations of personal gain, political bias or party policy, infinitely superior in such respects to, say, the scrutiny of proposals in a parliament, which in some other ways it resembled.

That is the role in which the eighteenth-century Enlightenment cast science, as the completest realization of the ethical ideals which underlie democracy. Scientists, so the story then ran, were

18. Feyerabend, *Against Method,* p. 188.

independent, self-sacrificing, internationally minded, devoted only to truth, hostile to any form of concealment or secrecy, indifferent to wealth and prestige. More recently, I have only once encountered this way of looking at science in its former glory; ironically, or perhaps pathetically, in the Medvedev papers, where that distinguished and much harassed Soviet biologist expresses the Enlightenment ideal in a heart-rendingly nostalgic form. "There is only one social group of people in the world," he tells us, "which, not only on account of its position in society but simply on account of the humane qualities inevitably inherent in it . . . is connected in a world-wide mutually dependent, mutually advantageous, mutually respecting system in friendship independent of national frontiers, constantly sharing among itself all possible help and interested to the utmost in the progress of mankind, of which it is the standard-bearer and motive force. This group consists. . . of the scientific, technical and culturally creative intelligentsia."[19] But such claims by no means go unchallenged nowadays. Scientists, so it is argued, are not disinterested critics; they adopt the point of view of the authorities in the society to which they belong, act, indeed, as their spokesmen. Science, Skolimowski has suggested, is once again on trial, Galileo is again before the courts but on quite different charges. "Science," so he writes, "is not tried as a force which attempts to upset the status quo, but as a force which represents the status quo. It is not tried as an emerging civilization, but as a part of a dying civilization."[20]

How did this new attitude come about? In part, because science now makes such tremendous demands for financial support; it is easy to represent it as one of many pressure groups. "Scientists," writes L. S. Feuer, "are becoming just one more of society's interest groups, lobbying for their greater share in the national income, and for the perquisites of power and prestige."[21] It was not a sensational but a respectable and by no means antiintellectual Australian journal of opinion which recently wrote:

19. Z. A. Medvedev, *The Medvedev Papers,* trans. Vera Rich (London: Macmillan, 1971), p. 170-71.

20. Henryk Skolimowski, "Science and the Modern Predicament," *New Scientist,* 24 February 1972, p. 437.

21. Feuer, *Scientific Intellectual,* p. 394.

> You shouldn't take all those predictions of an imminent ice age too seriously. By playing upon mankind's current obsession with doom and disaster, climate academics have averted what, to them, would have been an equally serious freeze: on the flow of money for meteorological research.
>
> It is a neat coincidence that the world's peril should have been discovered just in time to keep the funds coming in.[22]

Secondly, scientists have stepped into the political sphere as advisers to governments, sometimes in an attempt to justify public expenditure on science by demonstrating its practical usefulness, sometimes in the sincere belief that governments now *need* scientific advice. In the early months of 1972 the Australian Academy of Science produced two reports: one on the Concorde, one on DDT. In both cases it was accused, to its indignation, of whitewashing the government. The accusation is no doubt unjustified. What is of interest, however, is that it was not merely leveled, but widely believed. The Academy did not exist in the 1930s. But I do not think it would then have occurred to anybody to suspect that a report issued by an independent scientific body was motivated by a desire to stand in well with the government. No doubt, there is now abroad a widespread and depressing cynicism, a refusal to believe in the very possibility of disinterestedness, a cynicism flowing, as often as not, from an uneasy conscience. But the more intimate involvement of science with governments and industries has left it particularly vulnerable to the charge that it is merely their spokesman.

In and through its relationship with government and industry, science has lost, too, its old reputation as the exemplar of the ideal of open, public discussion. Much of its work—and work, often enough, on matters of great public concern—is now secret. That, so it is sometimes argued, is not a serious departure from tradition; scientists working within large corporations and government research institutes form by themselves a scientific community within which scientific proposals are subjected to proper scientific criticism. One may be permitted to doubt whether this is in fact so. Scientists are not always the best judges of what other scientists they ought to consult— the idea of consulting a biologist does not

22. *National Times,* Sydney, 31 March 1975.

come naturally to a physicist—and government departments, for example defense departments, do not normally cover the entire range of scientific activity. But that point aside, to abandon the ideal of public discussion is to abandon not only a methodological principle, which the size of the governmental scientific community might partly preserve, that the truth is more likely to be discovered in an atmosphere of mutual criticism; it is to abandon what was a key point in the morality of scientists.

I do not wish to deny, of course, that governments must sometimes work in secrecy. But the powerful are enamoured of secrecy beyond tolerable limits. For one thing, as Plato long ago pointed out, they are frightened men. Their instinct is to mark their requisitions for paper clips "highly confidential." For might not some spy deduce from a higher consumption of paper clips that morale is low, as they are neurotically twisted into uselessness? The paranoia of power knows no limits. Secrecy, furthermore, is one of the principal perquisites of the powerful: the feeling that they *really* know, in an ignorant world. As well, it protects them from criticism, it allows them to denounce their critics as ill informed—having taken great care to ensure that they will not be well informed. To live in this atmosphere is not consonant with the maintenance of the highest levels of free criticism.

It is relatively easy, indeed, to build up a picture of the scientist as a mercenary, prepared to serve any master who will supply him with his equipment. Many people were shocked when the Soviet Union proved to be so competent in the field of space technology for they had presumed that political freedom and advanced physical science go hand in hand. It is certainly no compliment to physical science that it flourishes in a Soviet Union in which creative literature survives only as an underground activity, and art is undisguised propaganda. One remembers George Orwell commenting on the Bernal-type enthusiasm of scientists for the Soviet Union: "They appear to think that the destruction of liberty is of no importance so long as their own line of work is for the moment unaffected."[23]

Or this complacent utterance by the social scientist George

23. George Orwell, "The Prevention of Literature," *Polemic* 2 (January 1946), reprinted in *The Collected Essays, Journalism and Letters of George Orwell*, ed. Sonia Orwell and Ian Angus, vol. 4 (London: Secker and Warburg, 1968), p. 70.

Lundberg, in his *Can Science Save Us:* "Physical scientists are, as a class, less likely to be disturbed than social scientists when a political upheaval comes along, because [their work] is recognised as of equal importance under any regime. Social science should strive for a similar position. . . . The services of *real* social scientists would be as indispensable to Fascists as to Communists and Democrats, just as are the services of physicists and physicians."[24] Such an interpretation of the scientist's social role has done a great deal to damage it—and not surprisingly. But the proper name for it is servility, not objectivity.

These considerations matter for our present purposes, however, only in so far as they affect the internal structure of science. They certainly affect it in so far as scientific "observations" may not be subject to adequate criticism when it is either politically dangerous to criticize them, or the range of critics is artificially limited by the demands of secrecy, or the critics are prepared to fake their results, or their criticisms arise out of an attempt to cut off funds for a line of research which might compete with their own. One cannot deny that all of these are, nowadays, real possibilities. When the *New Scientist* heads an article "Dirty Tricks in Science" or asks its readers to send it examples of faked experiments, we are moving in an atmosphere destructive of confidence in the morality of scientists and, hence, in the disinterestedness of their criticisms. And yet however greatly one regrets the disappearance of the old concept of the scientist as the prototype of personal integrity, one can still insist that frauds are more likely to be detected within science than outside it; that science is more genuinely international, and more open, than any other institution; that its methods of criticism ensure, by their thoroughness, a degree of objectivity which is not to be matched outside it. If there has been a decline, it does not constitute a fall. The abolition of science as we know it—whether in order to substitute a "proletarian" science in which the question is not "Is that likely to be true?" but "Is that doctrine in the interests of the party?" or in order to substitute for science the free play of personal speculation—is certainly not going to improve the level of objectivity, whatever can be said for it on other grounds.

24. George Lundberg, *Can Science Save Us* (New York: Longmans, Green, 1947), pp. 47–48.

The scientific community, one must admit, now contains far too many men and women for whom science is "just a career"; its postwar growth, like the postwar growth of universities, has been in many ways a tragedy for the world of learning. There is a real risk that the "human ants" will gradually take over positions of responsibility from scientists with broader vision and sharper critical capacities. But nothing we have so far said in criticism of science suggests that the collapse of science, should it eventuate, would be anything but a calamity.

I began by suggesting, however, that for many antiscientists objectivity is not so much impossible as reprehensible. Bacon and his disciples, Roszak tries to persuade us, went in search of "a philosophy of alienation." They "broke faith" with their environment, by establishing between it and themselves "the alienative dichotomy called 'objectivity.'" By that means they sought to increase their power, with nothing—no sensitivity to other people's feelings, no sensitivity to the life around them—to bar their access to "the delicate mysteries of man and nature." The cult of "objectivity," as much as their passion for analytic methods has led not only scientists, on this view, but, under their influence, all of us, to think of everything around us, people and biosphere, as "mere things on which we exercise power"—as nothing more than "behavioral surfaces." By way of illustration, Roszak quotes the psychologist Clark Hull, recommending a "prophylaxis" against anthropomorphic subjectivism—"to regard, from time to time, the behaving organism as a completely self-manipulating robot, constructed of materials as unlike ourselves as may be."[25] Objectivity, it is often suggested in this same spirit, is in practice a cloak for callousness, with its archetype the doctors in Hitler's concentration camps. Or perhaps Fermi's reply when Heisenberg suggested that it might be biologically dangerous to explode the hydrogen bomb: "But it is such a beautiful experiment."

Callousness, however, is not a peculiarity of scientists; it is encountered wherever there is absolute dedication to a form of activity. One finds it exhibited by mystics, revolutionaries, artists, even by lovers—in their relationship to third parties. To abolish science would not be to destroy it, but only to change its venue. It

25. Roszak, *Where the Wasteland Ends*, p. 169.

has nothing to do with objectivity; it is a product of "specialism," in the widest sense of that word. One does not demonstrate one's objectivity by refusing to look more broadly at the wider situation, to take into account the consequences, for others, of what one is doing.

Neither does one display objectivity by refusing to admit that people have sensitivities, anxieties. The assumption that it is impossible to be objective *about* feelings is quite baseless. To say that feelings are "subjective" is only to say that they *characterize subjects* or, at most, that they vary under similar circumstances from person to person. And there is no reason why we cannot look objectively at subjects or at differences in individual reactions; we do so in everyday life, novelists do so, psychologists have done so. To defend "objectivity," then, is not to defend either callousness or a psychology which confines itself to "behavioral surfaces"—at least insofar as such a psychology professes to be the *only* objective psychology. (In a particular inquiry, there may be no reason why it should pay attention to anything else.)

Roszak would retort, however, that there is a necessary connection between the kind of objectivity we took to be properly characteristic of science—the attempt to arrive at "public knowledge," which is impersonally presented for criticism, however personally derived—and the callousness which he and I would join in denouncing, that the one necessarily leads to the other. Bacon's admiration for Machiavelli is, he suggests, typical; the scientist sets out to destroy human communion by an attempt "to suppress the person and to treat that which comes to hand objectively—like a mere thing lacking in sacred autonomy."[26]

There is certainly a sense in which the quest for objective knowledge is an attempt to "suppress the person"; the scientist, when he seeks to test the "observations" which some other scientist has presented for his consideration, takes no account whatsoever of that scientist's hopes or fears, his expectations of glory or his need to impress his superiors. And insofar as he has "internalized" the scientific method of criticism, he tries to discount these factors even in himself—just as an artist may try to internalize the attitudes of those critics whose judgments he respects. Impersonality is in such contexts a virtue; the business of criticism, scientific or artistic or

26. Ibid., p. 170.

philosophical, is with the work, not the man, just as the business of an umpire is with the rules and impersonality is precisely what we demand from him. I know of no evidence that this impersonality spills over into situations in which it is not a virtue—except insofar as, I have already suggested, any sort of dedication may tend to do so. In itself it is of immense value; at worst, it has the effect, so we suggested, that there are certain social roles which it would be unwise to entrust to the scientists' care.

Certainly we ought all of us to be disturbed—even if we cannot accept the view that human beings possess a "sacred autonomy"—by the growing tendency to treat people as if they were merely things to be ordered as the planner pleases. And undoubtedly, modern inventions often encourage that tendency; nowhere do we feel as much like a thing as on a freeway, in an airport, or as a patient in a large modern hospital, designed as all three of them are for programmed robots rather than for people, people with their anxieties, insecurities, indecisions, incompetencies. It must certainly be admitted, too, that a startlingly large number of modern inventions have the effect of reducing the number of casual face-to-face relationships—the supermarket as compared with the corner shop, the private car as compared with the bus or train, the television or drive-in cinema as compared with the theatre, the high-rise building as compared with closely packed houses. The escalator which makes it impossible to exchange even a few words with a friend who passes us, moving up when we are moving down, is a symbol of our times.

Most of these, however, are practitioner inventions or bureaucratic innovations rather than the direct responsibility of science. I mentioned earlier that the totalitarian state will make use of art, science, philosophy, and religion as much as science in order to strengthen its power. So will commerce make use of all of them, so far as it can, to increase its profits. If we want to understand why there is now so strong a demand for innovations which in their general effect cut off human beings from direct contact with one another, we have to look beyond science to much broader social tendencies which are still little comprehended; they are related perhaps to the rapid growth of population. Here in turn the indirect social effects of science are of course important, but the chain of consequences is by no means direct.

What of the inhumanity of so much recent art? No doubt,

"functionalist" architecture, let us say, likes to think of itself as frozen geometry and therefore as peculiarly suitable for an aristo-scientific society, just as Mondrian argued that in a scientific age it is quite wrong for art to represent, as distinct from geometrizing, nature. Such a view, that it is always desirable to take science as one's exemplar even when the conception of public knowledge is totally inapplicable, is certainly an absurdity. Modern architecture in no way adds to our *knowledge* of geometry; neither do Mondrian's paintings, although he thought otherwise, reveal the secrets of nature to us. Once again, these particular absurdities could not have come into being did science not exist. But to say this is by no means to criticize science itself or to condemn objectivity as an enemy of art.

Critics of objectivity are sometimes critics of quantification. It is very natural to suppose that in order to be objective one must quantify. And the prevalence of this doctrine can certainly lead, as antiscientists argue, to an undue emphasis on what scientists know how to quantify and a corresponding neglect of what they do not know how to quantify, an emphasis and a neglect which contemporary talk about "the quality of life" is designed to rectify.[27] If I call the identification of the objective and the quantifiable "natural," this is because one important way in which the scientist can satisfy the requirement that he so express himself as to facilitate criticisms of his "observations" is to express them in a quantified form. The scientist will not write, let us say, "I heated the copper oxide until it became very hot" because "very hot" is at once vague and "subjective," that is, one person can call something "very hot" which the other will call "hot" without its being possible to conclude that either of the two is mistaken. (Compare the Japanese and the European concept of a "hot bath.") The scientist will write, rather, "I heated the copper oxide to a temperature of 120°C" or, more likely, and this time in the supposed interests of a *stupid* concept of objectivity, "the copper oxide *was heated* to a temperature of 120°C." In such instances the substitution for "hot" of a reference to a precise position on a temperature scale—the scale itself being determined by practical convenience—is entirely sensible and in no way misleading.

27. For more on this point see John Passmore, "Philosophy, Technology and the Quality of Life," in *Philosophy—the American Way*, ed. Peter Caws, in press.

But suppose, to take an example from within scientific institutions themselves, a committee responsible for distributing grants has to decide whether a certain scientist is more worthy than another scientist of a grant. Obviously this can be a difficult decision, one about which there can be serious disagreement—not in this case the sort of disagreement in which the parties can all be right. So there may be a temptation to look for an "objective test," perhaps that one of the scientists in question has written *x* papers which have been cited *xy* times, whereas the other scientist has written fewer papers, or papers which have been cited on fewer occasions. It does not take much reflection to see that these are bad criteria; the papers may be trivial and repetitious, the number of citations may be relatively high either because there are few papers in the scientist's field, so that what papers there are come to be mentioned in almost every bibliography, or because the scientist has been a very marginal participant in a number of many-authored papers produced by a rather indulgent research team.[28] On the other hand, a scientist may be working in a field where the competition is exceptionally intense, the literature very large, and to be mentioned at all is a considerable achievement. In the end one has to come back to a judgment on the candidate's work.

It is even more obvious that one is unlikely to find a quantitative substitute for the phrase "an extremely promising project"; highly distinguished scientists sometimes express opinions on this matter which range from total condemnation to the warmest enthusiasm. To substitute the phrase "a project of the sort which *x* scientists have undertaken and *x-3* scientists have succeeded in achieving results" would obviously be to prefer the unimaginative to the imaginative project. Whenever we meet with a quantified expression which professes to be a satisfactory substitute for a qualifying expression, we ought to ask whether there is any acceptable theory justifying this substitution or whether what is happening is just that something easy to quantify is replacing something difficult to quantify, in the name of "objectivity." Quantification is a blessing, but not as some do quantify. There are important things which have to be said but which we are not able to quantify; to refuse to take them seriously just for that reason is intellectually crippling.

28. See on this theme Henry W. Menard, *Science: Growth and Change* (Cambridge: Harvard University Press, 1971).

To say this is not, however, to denigrate either objectivity or quantification.

The general outcome of the argument of the last two chapters is that many of the major charges which have been brought against science cannot be sustained. It does not destroy uniqueness; it is not hostile to the imagination; it does not falsify by being abstract; it is as objective as the human condition permits and is not to be condemned for that reason. But I have not pretended that all is for the best in the best of all possible worlds, or denied that science has encouraged, even when it did not generate, attitudes of mind which can have adverse consequences. We are all of us, I think, coming to recognize that fact, scientists along with the rest. Science needed an injection of humility, and has had it; phrases like "dominion over nature," "the conquest of nature," now ring hollow in liberal-democratic ears—if not, I fear, in the ears of the developing, or the Communist, countries. We no longer take it for granted that we can "trust the experts." When it touches on human affairs, science is no longer accorded automatic respect. That specialization has its defects has become more and more apparent; indeed, we are now suffering, by way of reaction, from badly conceived interdisciplinary exercises. We are coming again to recognize—or am I being too optimistic?—that important discoveries do not necessarily originate in laboratories, that we can sometimes learn more from the historian than the sociologist, from the novelist or the film maker than from the experimental psychologist. Perhaps, even, we are beginning once more to look at nature with a contemplative eye, instead of always with the acquisitive gaze of a "developer" or the transformative eyes of a technologist.

But as Ravetz has argued, there is still need for the establishment of centers for "critical science."[29] Scientists themselves are coming to be conscious of that fact; such centers are already in being in a number of different places, even if not under that name. For a time it looked as if science might be an exception to the proverbial wisdom that "every blessing brings a curse" or to Ecclesiastes' "he that increaseth knowledge increaseth sorrow." But it has turned out not to be such an exception. If only for that reason we need to

29. Ravetz, *Scientific Knowledge and Its Social Problems*, pp. 421–34.

look more carefully at science, its structure, its institutions, its assumptions, its economic and social effects, than we have customarily done.

That task cannot safely be left to scientists, working alone. Ravetz's own exemplar of a "critical scientist" is Barry Commoner. But in Commoner's case we can readily see the kind of weakness which the scientist moving into "critical science" is likely to take with him. Characteristically, Commoner tries to enunciate "laws":[30] that "Nature knows best," that "everything is connected with everything else," that "you can't do one thing at a time," "laws" which, until recently, were to be found enumerated only in romantic antiscientific literature and romantic antiscientific philosophy. The first "law" is as it stands absurd: Nature neither knows nor cares; only human beings can do either of these things. All that can properly be said is that we must not assume that every technological innovation is superior to the natural processes it displaces, that every technological innovation is an improvement. "The cardinal tendency of progress," J. D. Bernal once wrote, "is the replacement of an indifferent chance environment by a deliberately created one."[31] Against this attitude, one has certainly to insist that an "indifferent chance environment" is often—not *always*, as Commoner's "law" would suggest—a much better place for human beings to live in than the product of their deliberate creation, a straggling village than a geometrically designed "high-rise," a diversified countryside of swamps and hills than a leveled-off park. For the "indifferent chance environment" has grown up as a result of a long series of intricate adjustments; the technologically created environment tends to be the product of an obsession with a single objective, or a very limited set of objectives. As Commoner himself emphasizes, too, the presumption that a man-made product can safely be introduced, as an "improvement," into a natural system can easily be a calamitous one. "Nature knows best," nevertheless, is a highly misleading way of saying what Commoner wants to say.

To make such points is not to deny that Commoner has contrib-

30. Barry Commoner, *The Closing Circle* (New York: Alfred A. Knopf, 1971), p. 33 ff.
31. J. D. Bernal, *The World, the Flesh, and the Devil* (Bloomington: Indiana University Press, 1969), p. 66.

uted to the emergence of "critical science." *Anyone* coming into this area from a special field is likely to suffer from the defects of his training; the philosopher of science, let us say, from too close an attention to logical relationships, too little sense of science as a living activity, the economist from too slight an appreciation of the waywardness of scientific discovery. It is only to insist that critical science ought not to be left to scientists. Science, like education, is so central to our culture that a great many forms of expertise can be brought to bear on it. But as the analogy with education suggests, there is a considerable risk that just for that very reason critical science will not prove attractive to minds of the highest order, that it will attract those who prefer to work between disciplines because they have not fully mastered any particular discipline, or that their expertise will somehow not bear upon the central, but only on peripheral, problems. For this is certainly what has happened in the case of education.

For my part, the question which still leaves me most uneasy is how to reconcile my concern for freedom of inquiry with a real fear of the consequences of a burgeoning technology. There are those who would argue that the villain of earlier historians of science, Cardinal Bellarmine, had right on his side when he urged upon Galileo the need to consider the social consequences of promulgating his astronomical views, the likelihood that he would disrupt the fabric of society, even if it is equally true that Galileo had right on his side when he stood by his discoveries. So the elements were already present of a Hegelian-style tragedy, the conflict of Right with Right, which the growth of technology has once again brought vividly to our attention.[32] But many of us who would be prepared to reply to Bellarmine that the only good society is a constantly questioning society are not so confident in our attitude to a science which lends itself, through the innovations it makes possible, to the destruction of life and liberty.

To close down science, to withdraw funds for its support, just because it *might* have such effects would be at once to sacrifice one of mankind's greatest achievements and to infringe the liberty we pretended to be preserving. (That is even leaving aside the fact that

32. See for a discussion of this line of reasoning Jerome R. Ravetz, "Tragedy in the History of Science," in *Changing Perspectives in the History of Science: Essays in Honour of Joseph Needham* (London: Heinemann, 1973), pp. 204–22.

we need more science to mitigate the present effects of technology.) And short of closing it down, I have also argued, we can never wholly avoid the risk of adverse consequences. But that still leaves open a class of cases in which we can *foresee* serious risks. It is at that point that even the most enthusiastic defender of science has problems to face which he cannot properly thrust aside as no business of his. It is obvious that we should do what we can to reduce such risks. But can we carry that policy so far as either to prevent by law the continuance of certain investigations or to withdraw funds from them when they are of very considerable scientific interest?

To those who agree with Thomas Gray that "where ignorance is bliss 'Tis folly to be wise," or like to remind us that the Sirens attracted Odysseus and his men with promises of greater knowledge, Fermi answers uncompromisingly: "Whatever Nature has in store for mankind, unpleasant as it may be, men must accept, for ignorance is never better than knowledge."[33] That answer has the virtue of simplicity. And so does the alternative doctrine that all science must be subordinated to and serve "the public good." But we can agree that inquiry is a good without agreeing that "anything goes" when it is a question of synthesizing a new chemical substance or modifying viruses or setting up new sources of powerful radiation. And we can agree that a grants committee, let us say, can reasonably take into account the social importance of a line of enquiry when it is distributing funds without believing either that there is something describable as "*the* public good" or that *only* social importance ought to be taken into account.

To adopt this position has the effect, which some would regard as intolerable, that we have no unyielding principle on which to base our decisions, neither the principle that nothing is of any consequence except that new knowledge should be acquired, nor the principle that new knowledge matters not at all except insofar as it contributes to "public good"—or is "socially relevant" or what you will. But unless we are fanatics, that is the kind of situation we commonly confront when we have to make a funda-

33. Enrico Fermi, cited in Feuer, *Scientific Intellectual,* p. 396. Many of the points at issue are well discussed by Max Black in a lecture entitled "Is Scientific Neutrality a Myth?" delivered at the annual meeting of the American Association for the Advancement of Science, New York, 27 January 1975.

mental political decision. Opposite principles have to be weighed, and there is no mechanical method of weighing them.

Whatever our decision on particular questions, we must certainly never forget, in the flurry of reappraisal, that aristoscience has brought us great gifts, both intellectual and practical. It has made plain how much can be achieved by intellectual cooperation, by imagination and hard work in close alliance, by high standards of honesty, directness, and concern for truth. "Scientists," C. S. Peirce once wrote, "are the best of men." No scientist myself, I am still inclined to think that this is true—without, in claiming as much, at all denying that individual scientists can well be, in Mishan's words, "petty, vicious, coarse-grained, paranoid even" or that *The Double Helix* is an accurate enough picture of the extremes of scientific competitiveness.[34] The best of men are far from being perfect. Insofar as the revolt against science condemns science for making of itself the instrument of power, looks with dismay on the devastation to which science-based technology has given rise, rejects a world made grey by standardization or a world in which the individual counts for less and less, and seeks to reinstate the imagination and direct sensual enjoyment, one can sympathize with its motives even when one believes that its accusations rest on a misunderstanding of science. But the attack on disciplined thinking, the revival of occultism with its doctrine of "hidden truths" to be revealed by magical means, the demand for instant gratification in every area of human life, the rejection of the idea of learning, of discipleship, of tradition—these, I freely confess, fill me with horror and dismay. And it is not only science which suffers but all the activities which particularly distinguish human society, as compared with a society of bees and ants—art, history, philosophy, social innovation, technological achievement, the spirit of critical inquiry.

34. For a scientist's comment on *The Double Helix*, see P. B. Medawar's chapter "Lucky Jim" in his *Hope of Progress* (London: Methuen, 1972), together with his comments in the preface, pp. 11–12.

DATE
DISCHARGED
DISCHARGED
DISCHARGED